Earthly Adams and Pious Philosophers

Medieval and Early Modern Philosophy and Science

VOLUME 43

The titles published in this series are listed at *brill.com/memps*

Earthly Adams and Pious Philosophers

A Theological Anthropological Lens to the Sixteenth-Century Astronomical Revolution

By

Nienke Roelants

BRILL

LEIDEN | BOSTON

Cover illustration: Detail from *Adam and Eve*, by Lucas Cranach the Elder, 1528. Inventory 1890, no. 1459, Galleria degli Uffizi, Florence.

The Library of Congress Cataloging-in-Publication Data is available online at http://catalog.loc.gov
LC record available at https://lccn.loc.gov/2024046010

Typeface for the Latin, Greek, and Cyrillic scripts: "Brill". See and download: brill.com/brill-typeface.

ISSN 2468-6808
ISBN 978-90-04-34901-8 (hardback)
ISBN 978-90-04-38775-1 (e-book)
DOI 10.1163/9789004387751

This book is printed on acid-free paper and produced in a sustainable manner.

Contents

Acknowledgements

This book has come a long way, and could not have been realized without the help of many. Originally a Ph.D. dissertation, it is the result of several years of work at Ghent University under the supervision of Steven Vanden Broecke, who provided invaluable expertise in the field of early modern intellectual history and the history of science, and continues to do so. Financial support for the research project was provided by the Research Foundation—Flanders (FWO). Others from the then Ghent-based academic community to whom I'm grateful for their encouragement and for helping to shape ideas by providing feedback on papers and presentations were Eric Schliesser and Maarten Van Dyck. I would also like to thank the members of the doctoral jury, Patrick Boner, Kathleen Crowther, and Adam Mosley, for their expert advice and fresh perspectives.

Between dissertation and book came several years of other professional activities and, well, life. However, my wish and intention to publish my dissertation as a book didn't fade. I'm grateful to Christoph Lüthy, the series editor of Medieval and Early Modern Philosophy and Science, for being able to realize this ambition in such a highly reputed book series. I would also like to thank the peer reviewers for their indispensable advice and the editorial team at Brill for making this publication possible.

My final and deepest thanks, of course, go to my dear family, to Peter, Felix, and Abke, to whom I dedicate this book.

INTRODUCTION

On Earthly Adams and Pious Philosophers

If a Dictionary of Scientific Biography had been compiled by the Renaissance astronomers with whom we will concern ourselves in this present volume, the entry on Adam would credit the first human for being the founder of the art of astronomy, or of any art for that matter. Early modern astronomers, Protestant and Catholic alike, traced back the origin of their art to the first parent, the original progenitor. They subsequently created a chronological succession of astronomers, from the earliest of times extending to their own days. Such prefatory histories of astronomy were found in dedicatory letters, prefaces, prolegomena to astronomical treatises, and inaugural lectures at universities.[1] The first father was perceived, of course, as a historical character, and so the sixteenth-century *modus operandi* of placing Adam at the head of a disciplinary genealogy underlines the then-unified view of the scientist and the scientist's religious persona. There was no secular understanding of the self nor of the world; anthropological and ontological views, microcosm and macrocosm, by definition were theologically informed. The Genesis account was appropriated, both by theologians and laymen, in order to help them understand, shape, and give meaning to the contemporary relationships between God and humanity, and between human beings themselves. The (hi)story of Adam and Eve, being omnipresent in sermons, in Biblical commentaries, in paintings and in medical textbooks, played a decisive role in the formation of sixteenth-century Lutheran confessional culture and identity.[2] This can be seen in the works of the Lutheran "Wittenbergian" Renaissance astronomers Georg Joachim Rheticus (1514–1574), Caspar Peucer (1525–1602), and Tycho Brahe (1546–1601), who were occupied with restoring the age-old art of astronomy, a concern that had arisen among their late-fifteenth century predecessors.[3] Their respective representations of their scholarly selves, their art, their object of study, and their intellectual opponents within the context of the sixteenth-century astronomical revolution document the presence and the meaning of the Adamic figure in learned culture, which in turn constitutes a gateway to a better understanding of the contemporaneous calls for a pious study of the world. Whilst featuring still in lay culture, Eve was close to absent from the learned picture.

1 Jardine, *Birth of History and Philosophy of Science*, 261–262.

2 Crowther, *Adam and Eve*, 5–6, 13–14.

3 On the history of Renaissance mathematics, see: Rose, *Italian Renaissance of Mathematics*; Swerdlow, "Science and Humanism," 131–168; Swerdlow, "Regiomontanus," 165–198.

 | DOI:10.1163/9789004387751_002

"Wittenbergian" refers to the sphere of influence of Wittenberg University. Owing to its exceptional situation as the *heimat* of the Reformation and its position at that time as an important center of astronomical activity, Wittenberg University provides an ideal lens through which to view the dynamics between religion and the sixteenth-century astronomical revolution. The Academia Wittebergensis was founded in 1502 by Elector of Saxony Frederick III the Wise (1462–1525) and was home to Reformer Martin Luther (1483–1546) who taught there until his death, having entered the university as a professor in moral philosophy in 1508.[4] Universities were seen as places from which the Reformation could become more firmly established and as a principal means to reform society. New universities were founded, while others—already established, like Wittenberg, Tübingen and Copenhagen—were redefined along Protestant lines.[5] Due to Wittenberg's status as a model for Reformed universities and its strong textbook tradition, Luther's theological legacy and the written output of Wittenberg scholar Philipp Melanchthon (1497–1560) were institutionalized and consolidated at other Reformed universities. Even beyond the borders of the Holy Roman Empire, Reformers such as those in Geneva and the Huguenots in France paid close attention to what was being taught at Wittenberg, which demonstrates the importance of its university and its educational policy for European Reformed culture as a whole.[6] Two of our three named astronomers—Rheticus and Peucer—were educated and professionally active at Wittenberg University, while Tycho's education at Copenhagen University would have strongly corresponded to the learning propagated at Wittenberg.

Debates about the (geo)heliocentric set-up of the universe, the existence of solid celestial spheres, and the particulars of celestial matter were initiated and sustained by Rheticus, Tycho, and Peucer. In so doing, they contributed to the sixteenth-century astronomical revolution that prompted the early modern transformation of the Aristotelian-Ptolemaic world picture.[7] Their writings document two key episodes of this revolution—namely, the earliest reception of the heliocentric astronomy of Nicolaus Copernicus (1473–1543) in the 1540s and 1550s, and the debates ensuing from the observations of celestial novelties in the 1570s and 1580s. Their accounts offer us a demonstration of the extent to which a canonical cosmology had been eroded, and the Scholastic hierarchy of sciences undermined. The confidence which allowed mathematicians to cross disciplinary boundaries into the domain of natural philosophy and

4 Junghans, "Luther's Wittenberg," 22; Kathe, *Die Wittenberger Philosophische Fakultät*, 1.

5 Austin, "Academic Exchanges," 39–40, 51.

6 Bellucci, *Science de la nature*, 7.

7 Barker, "Lutheran Contribution," 57; Cabezón, "Introduction," 4.

to begin to act as theorizing astronomers gave rise to a phenomenon known as "disciplinary role-roversal."[8] In the context of the celestial novelties which occurred during the 1570s and 1580s, observations of which inspired a select few astronomers to leave canonical cosmology, Patrick Boner and Dario Tessicini speak of "the dawn of the astronomer-philosopher."[9]

The changing relationship between the mathematical disciplines and natural philosophy was just one manifestation of the late-sixteenth century reassessment of the foundations of knowledge at the expense of the credibility of Scholastic natural philosophy. Holy Scripture and the Book of Nature alike were taken seriously as cognitive categories.[10] Rheticus, Tycho, and Peucer articulated their respective reassessments of the foundations of knowledge in terms of theological anthropology. When not to be dismissed as mere rhetoric used in order to please theological authority, royal patronage or other readership, the distinctly religious slant of their pronouncements—as conveyed in university textbooks, scientific treatises, academic declamations *ex officio*, and correspondence among peers—is revealing of the ways in which they have constructed public personae as learned men dealing with God's creation after Adam's Fall, tailored to the markedly pessimist anthropology of Reformer Martin Luther. Insofar as Philipp Melanchthon had developed a Lutheran natural philosophy based on the principle of the distrust of human reason, these astronomers echoed a pessimist epistemology in cosmological debate.[11] Hence, their astronomical-philosophical writings show that Luther's anthropology and Melanchthon's Lutheran natural philosophy had crystallized into broader learned discourse at Wittenberg and in scientific culture.

Theological anthropology is a fruitful category for us to explore because it explains the contemporaneous references to living a life after the Fall, and to the calls for piety. It refers to the systematic theological account of human nature, determined by the understanding of the impact of key events in eschatological history like the Fall or Redemption, and crystallized in doctrines about things such as creation, original sin, and grace.[12] Peter Harrison introduced this particular angle to the early modern history of science in his work *The Fall of Man and the Foundations of Science*. He convincingly describes how, in seventeenth-century England, the emphasis of Reformed theology on the

8 Westman, *Copernican Question*, 230. For the putative shift in the astronomer's role, see: idem, "Astronomer's Role," 104–147.

9 Boner and Tessicini, "Introduction," IX.

10 Pomata, "Observation Rising," 45.

11 Kusukawa, *Transformation of Natural Philosophy*.

12 Janz, *Luther*, 1.

pervasive effects of original sin on the human condition informed contemporaneous discussions about the foundations of knowledge. For seventeenth-century scientists, the theological understanding of the story of the Fall provided them with a specific appraisal and articulation of their own intellectual abilities. Their aim was to restore Adamic knowledge.[13] Adding a theological anthropological dimension to the Renaissance reassessment of the foundations of knowledge will allow the integration into a single frame of reference of some of the major shifts that occurred in the late-sixteenth century study of the heavens. It considerably deepens our current understanding, for example, of the widely aired demand for a pious philosophy, the emergence of Mosaic philosophies, and the ubiquity of the metaphor of the Book of Nature.[14] The rules of intellectual piety were argued to be changing and became tailored to a Reformed understanding of postlapsarian anthropology. In this period, the terms "pious philosophy" and "Christian philosophy" evolved into standard expressions to designate philosophies opposed to Aristotelianism.[15] Harnessing the powers of persuasive rhetoric, Reformers and astronomers advocating novel or non-canonical cosmological claims ran through the same process of distancing themselves from the Scholastic lens applied to their objects of study. For cosmology, the written polemic took place outside the university context: at Wittenberg, a canonical approach was preferred.

Besides our three named astronomers, Wittenberg mathematician Erasmus Reinhold (1511–1553) will also command some of our attention in this study, but to a lesser extent, given that he did not—at least not in writing—adopt the guise of a theorizing astronomer in cosmological debate. Alongside Rheticus and Peucer, Reinhold was identified as a third member of the Melanchthon circle, referring to the intellectual custody of Philipp Melanchthon.[16] From a religion-science dynamics perspective, John Christianson had previously grouped together Reinhold, Peucer, and Tycho as the Lutheran school of geostatic Copernican studies.[17] If all four men can be considered a group on the basis of their religious identity, this is not, however, their only point of convergence. Another common denominator is their pre-Keplerian understand-

13 Harrison, *Fall of Man*.

14 Blair, "Mosaic Physics," 32–58; Crowther-Heyck, "Wonderful Secrets of Nature," 253–273; Jorink, *Reading the Book of Nature*; Mandelbrote and Bennett, *Garden*; van Berkel and Vanderjagt, *Book of Nature*; van der Meer and Mandelbrote, *Nature and Scripture in the Abrahamic Religions*.

15 Blair, "Mosaic Physics," 32, 34; Menn, "Intellectual Setting," 1: 33–86.

16 Westman, "Melanchthon Circle," 165–193.

17 Christianson, "Copernicus and the Lutherans," 1–10; Tredwell and Barker, "Copernicus' First Friends," 143–166.

ing of the ontological potential of astronomical models. Tübingen astronomer Michael Maestlin (1550–1630) was quite right in commending his former student, the Lutheran astronomer Johannes Kepler (1571–1630), for having demonstrated *a priori* what his predecessors had achieved only *a posteriori*: namely, the reality of the heliocentric system, through which could be revealed the secret of God's creation. Kepler's explicit *a priori* perspective in his *Mysterium cosmographicum* (1596) was unprecedented: astronomers traditionally confined themselves to descriptions of the mere effects of the celestial bodies perceived from earth, to "saving the phenomena." Pietro Daniel Omodeo describes this pre-Keplerian stage as an *a posteriori* astronomy.[18] This *a posteriori* knowability of the universe, however, was also understood as a theological anthropological condition by these astronomers, even by Rheticus, who is argued to be a full-fledged supporter of Copernicus's hypothesis as a true model of the universe. The Lutheran-Wittenbergian theological anthropological blueprint established here for cosmological debate awaits extrapolation not only into the seventeenth century but also to Wittenbergian astronomers at other contemporary Lutheran centers of astronomical activity, such as Christoph Rothmann (c. 1560-c. 1601) at the court of William IV (1532–1592), Landgrave of Hesse-Kassel.

The dynamics here brought into focus—between early Lutheranism and Renaissance astronomy—do not entail the specific attitudes of either Catholic or Protestant theologians towards Copernican astronomy.[19] Nor do they concentrate on astrology in sixteenth-century Lutheran circles, since this work has already been persuasively completed by the likes of Peter Barker, Miguel Granada, Sachiko Kusukawa, and Charlotte Methuen. All have underlined the importance of astrology as a vehicle for the study of astronomy: The better grasp of God's providential plan was a decisive motive for Lutheran astronomers to participate in the reformation of astronomy.[20] The present study complements earlier research on astrology but will not delve too deeply into astrology itself. Instead, we look at the impact of astronomical inquiry into the celes-

18 Omodeo, *Copernicus*; In particular, see chapter 6 "*A priori* and *a posteriori*: Two Approaches to Heliocentrism," 234–254.

19 Boas Hall, *Scientific Renaissance*, 125–127; Westman, "Copernicans and the Churches," 76–113. For recent appraisals of the attitudes of Lutheran and Catholic theologians towards Copernicus's astronomy see: Barker, "Lutheran Contribution," 34–37; Lerner, "Aux origines de la polémique anticopernicienne (I)," 681–721; idem, "Aux origines de la polémique anticopernicienne (II)," 410–424.

20 Barker, "Role of Religion," 61; Granada, "Tycho Brahe, Caspar Peucer," 563–583; Kusukawa, *Transformation of Natural Philosophy*, 4; Methuen, "Special Providence," 99–113; eadem, "Role of the Heavens," 385–403.

tial realm on the cosmological debate that was occurring at the same time, and thus put the spotlight on the disciplinary interplay between astronomy and natural philosophy. The crux is not one of the predictability of divine providence, but is rather an appraisal of the actual knowability of divine creation.

This is not a history of scientific progress. By privileging sixteenth-century Lutheran astronomy, I have no intention of arguing in favor of the exceptionalism of early modern Lutheran science, nor to put Protestant theology in the driving seat of a general early modern skeptical crisis. Gábor Almási has argued against the exceptionalism of Lutheran science by highlighting that similar religious motifs occurred in texts by scholars of other confessional denominations.[21] Rather than using Lutheran as an antithesis to Catholic, the Lutheran attribution derives from the strong parallels between Martin Luther's views and those of the astronomers, doctrines which may or may not have been different from the Catholic position. There is more to a confessional approach than exceptionalism. It allows for an integration into their proper denominational context of contemporaneous statements made, for example, about the true church, the Book of Nature, and the limited knowability of divine creation. A next step is to compare these findings to the sixteenth-century discussions dealing with changing criteria of scientific credibility in non-Lutheran confessional and intellectual settings from a theological anthropological perspective. Regarding the tendency to prioritize Protestant theology in discussions of the early modern skeptical crisis, Richard Popkin's influential argument in the 1960s was that Luther's dismissal of tradition as a criterion of certitude within theological discourse resulted in a larger debate regarding the vindication of knowledge. Due to its coincidence with the revival of ancient Greek skeptical literature in the second half of the sixteenth century, the so-called intellectual crisis of the Reformation "was to unleash a skeptical crisis not only in theology but also, shortly thereafter, in the sciences and all other areas of human knowledge."[22] I am more inclined to trace this anti-dogmatist attitude back to the mathematical Renaissance of the fifteenth century and—in line with Ian Maclean—to the medieval tradition of *libertas philosophandi*. In rethinking Popkin's skepticism thesis, Maclean argues for a long tradition of free inquiry stretching from the Middle Ages to the seventeenth century, characterized by critical thinking about epistemology, by a tendency to question authority, and closely associated with the independence of philosophy from theological domination. This skepticism is closely allied with different modes of thinking, such

21 Almási, "Rethinking Sixteenth-Century 'Lutheran Astronomy'," 5–20.

22 Popkin, *History of Scepticism*, 15. The early modern revival of skepticism is linked to the publication of Sextus Empiricus' (*fl.c.* 200 CE) *Outlines* in Geneva, 1562.

as epistemological pessimism.[23] The call for *libertas philosophandi* resonated through the early history of Copernicanism in both Protestant and Catholic pro-Copernican writings.[24] In the 1540 *Narratio prima*, Rheticus expounded Copernicus's theory, but only after having reminded his readership of a quotation by the ancient Platonist philosopher Alcinous (*fl.c.*150 CE). On the frontispiece, the quotation reads as follows: "It is necessary that the one who sets out to philosophize be free in thought." For early modern Catholic astronomy, the matter of freedom in thought has been researched mostly in relation to repression by ecclesiastical authorities. Obvious examples from the late sixteenth and seventeenth century are the punishments meted out to Giordano Bruno (1548–1600) and Galileo Galilei (1564–1642). The cosmological interventions of the Wittenbergian astronomers add another, earlier, chapter to the history of free thinking. The theorizing astronomers argued for freedom from authority or the freedom to think independently from canonic tradition. Their anti-dogmatist stance did not concern freedom from theology—far from it—, but indicated instead the astronomer's right to think freely from Aristotelian domination.

Chapter 1 underlines the just nature of the proposition regarding early modern scientists made by Amos Funkenstein that "never before or after were science, philosophy, and theology seen as almost one and the same occupation."[25] It takes us to the Arts Faculty at Wittenberg University—astronomy was part of the quadrivium—between the 1530s and 1550s as an institutional point of departure. The academic community represented its academia as a place of learning that protected the true church, indeed that itself embodied the true church. After discussing Luther's notion of the true church, the chapter expands upon the self-profiling of the Wittenberg scholarly community as the lineage of that true church, and the ensuing responsibilities of the learned in general and the astronomers in particular towards the religious community and before God. Inasmuch as academia and church were one and the same, there was no distinction between the faithful astronomer and the learned believer.

Chapter 2 exchanges the collective for the individual perspective and brings to light the particulars of Luther's anthropology and epistemology based on his lectures on the Book of Genesis. Whereas Philipp Melanchthon has often held center stage when analyzing Lutheran science, the spotlight is here shifted to

23 Maclean, "'Sceptical Crisis' Reconsidered," 247–274.
24 Granada, "Travail of Philosophical Liberty," 47–67.
25 Funkenstein, *Theology and the Scientific Imagination*, 3; Osler, "Mixing Metaphors," 101–103.

the theological anthropological views of Martin Luther. Our findings do tie in with earlier research by Sachiko Kusukawa and Dino Bellucci on the *praeceptor Germaniae* and Melanchthon's shaping of natural philosophy at the service of the Reformation.[26] The next four chapters demonstrate how Luther's negative and positive appraisals of postlapsarian reason resonated in learned astronomical discourse: Chapter 3 focuses on astronomy as a form of learned devotion. Rheticus, Peucer, Reinhold and Tycho are seen to have shared the Reformer's contemplative understanding of their astronomical activities. Chapters 4 to 6 highlight the theological anthropological pessimism present in the extra-academic cosmological contributions by Rheticus, Tycho and Brahe, a strand in their thinking which goes some way towards explaining their tendency to overstep prescribed disciplinary divisions. These scholars refrained from thinking about the universe as "epistemologically accessible."[27] Throughout the Wittenbergian episodes of the Lutheran contributions to the sixteenth-century astronomical revolution, epistemological pessimism emerges as a constitutive part of Lutheran intellectual piety.

26 Bellucci, *Science de la nature*, 12.

27 A paraphrasis of José Cabezón on the scholastic presumption of the epistemological accessibility or intelligibility of the world; see Cabezón, "Introduction," 5.

CHAPTER 1

The True Church at Wittenberg University

Martin Luther's dissent from the Roman Catholic hierarchy formally and irrevocably ended the religious-political unity of the Holy Roman Empire. Amidst the religious and political turmoil of their days, Luther and his followers carved out a confessional identity distinct from that of the papal church.[1] Key to this identity was the idea of a (virtual) true church, migrating its way through world history while striving for survival. The true church was perceived as a community of God's Word that started with the divine creation of Adam; from then onwards, the Word was passed down through successive generations.[2] Luther's discourse on the true church is derived from his commentary on Genesis, an exegesis which proved to be the first in a tradition of major Protestant biblical commentaries. The academic community at Wittenberg strongly identified itself with the true church, which was understood to be a learned church. The astronomers' membership of the true church determined their conception of the role of Wittenberg University as well as of their own responsibilities as (part of) the learned community. It shaped their Protestant academic identity.

Declamations *ex officio* issued between the 1530s and 1550s by Melanchthon and his circle on (natural) philosophy and the liberal arts, as well as prefaces in university textbooks on astronomy used at the Wittenberg Arts Faculty, all of them dedicated to secular authorities, ostensibly bear the imprint of the Reformed confessional identity. The humanist declamation had become a standard part of regular instruction from Melanchthon's curriculum reform of 1523 onwards, and was acknowledged as such in the statutes of Wittenberg University and of the Arts Faculty. As a result, there was a general increase in the use of demonstrative speeches. Epideictic rhetoric or demonstrative oratory was a mode of discourse in which praise or blame was heaped upon a person or discipline, and the value of treating a subject was defended. Three different moments gave rise to speeches in academic life: namely, the monthly oration held in turn by different members of the academy; the lectures inaugurating the establishment of an academic chair; and, finally, some other specific occasions. Such occasions could include the doctor and master promotions at the respective faculties, the reading aloud of the university statutes, and the succession of

1 Maxfield, *Luther's Lectures*, 169.

2 Pohlig, *Zwischen Gelehrsamkeit*, 294, 296.

 | DOI:10.1163/9789004387751_003

rectors and funerals.[3] The textbooks of Philipp Melanchthon on world history, dialectics, and natural philosophy turned Wittenberg into an influential place for textbook production.[4] Since his entering of the university as a professor in Greek, Melanchthon would teach both in the Arts and Theology Faculties for the rest of his life, and was an important author of textbooks and theological works.[5] This chapter illustrates the intense collaboration between both faculties. In Luther's view, a new university curriculum had to support the Reformed Bible-based theology. Scholastic universities, in his opinion, were concerned more with reading the *Libri quattor sententiarum* (c. 1150) by twelfth-century author Petrus Lombardus (c. 1096–1160) than with reading the Bible, an order of priority he wished to reverse.[6] It was Melanchthon who developed and implemented the subsequent curricular reforms at the Arts Faculty.

1 Luther's Notion of the True Church

Protestants defined themselves as the direct descendants of the true church and the stronghold of original Christian doctrine, with Adam as its first member.[7] Within the sixteenth-century context of the conflict with the papacy, the self-fashioning of the Reformers as the true church was key to the Protestant identity.[8] This identity was essentially antagonistic—and was developed in opposition—to the papal church. Luther traced back the existence of the Catholic and the Lutheran churches to the two first sons of Adam and Eve, Cain and Abel. These two brothers represented the split between the two different religious communities. Cain's church was characterized by pride, power, hypocrisy, and violence; Abel's was distinguished by humility and faith.[9] The church of the pope was identified with the church of the elder brother Cain; Luther and his followers identified themselves as the church of the true believer Abel.[10] This dualistic imagery of Cain and Abel became deeply rooted in Lutheran lay culture as well, and emerged as a motif in plays and other texts.[11]

---

3 Kathe, *Die Wittenberger Philosophische Fakultät*, 77; Müller, "Zur Chronologie," 117–118; Swerdlow, "Science and Humanism," 141.
4 Valleriani, Federau and Nicolaeva, "Hidden Praeceptor," 407–436.
5 Wengert, "Philip Melanchthon," 313.
6 Methuen, "History and Heresy," 10–11.
7 Ben-Tov, *Lutheran Humanists*, 39.
8 Maxfield, *Luther's Lectures*, 169.
9 Ibid., 157–158.
10 Ibid., 151.
11 Crowther, *Adam and Eve*, 226–257.

Contrary to Catholic doctrine, the church in Lutheran understanding was not an ecclesiastical institution but consisted of adherents to true doctrine, with a chronology that began with Adam in the Garden of Eden.[12] To Luther, the greatest of God's gifts to man was membership of the true church, members of which were named "children of God."[13] His argument was that Moses had thus designated the Patriarchs to whom the divine promise of salvation had been made: they were the first *filii Dei*.[14] According to Luther, the history of creation, and of God's Word in general, reminded its readers of the divine promise that the members of the true church would be forgiven and would eventually merit eternal salvation.[15] This promise was man's living fellowship with God.[16] In the Gospel it had materialized in the Atonement of Christ; his suffering on the cross—his first advent on earth—had assured redemption for humankind.[17] Luther came to integrate the divine promise of redemption within the history "of the creation fallen, redeemed and sighing for fulfillment."[18] The divine prospect of eternal salvation connected the people of the Old Testament with the New Testament and the living church, or—better still—it defined the Patriarchs as the earliest members of the true church that survived until the end of time. The promise of redemption was passed on from Adam to Seth, from Seth to Noah, and so on.[19] It heralded, in fact, Christ's second coming on earth, an event that announced the end of time and the general Resurrection.[20] Faith, from a Lutheran point of view, was the absolute trust in the salvation promised by God.[21]

The people of the Old Testament were no mere *figurae*.[22] As a result of a literal-historical reading of Scripture, the people of the Old Testament were perceived as an authentic people. Adam was a historical character. Biblical events like the Fall and the Deluge were understood and chronicled as milestones in world history. Genesis was a historical source about the first two thousand years

12 Ben-Tov, *Lutheran Humanists*, 32; Esnault, "Introduction," 11.
13 Luther, "Enarratio," 42: 270 (Commentary on Gen.6:1.2).
14 Ibid., 269–270 (Commentary on Gen.6:2).
15 Bayer, "Luther as an Interpreter," 76; Luther, "Enarratio," 42:132 (Commentary on Gen.3:12); and 269–270 (Commentary on Gen.6:2).
16 Baumgaertel, "Promise," 3.
17 Maxfield, *Luther's Lectures*, 35.
18 Bayer, "Luther as an Interpreter," 76, 79; Benin, *Footprints of God*, 186; Preus, *From Shadow*, 4.
19 Maxfield, *Luther's Lectures*, 35.
20 Barnes, *Prophecy*, 32.
21 Kusukawa, *Transformation of Natural Philosophy*, 160.
22 Preus, *From Shadow*, 6.

of history. God's creation of the world, the Fall of Adam and Eve, the universal Deluge, and other Biblical events were seen as historical facts: men and women of Luther's day were contemporary actors in a universal history that commenced almost six millennia earlier in the Garden of Eden. Luther propagated the earliest church as authoritative, as the true model. The age of the Patriarchs was portrayed as a golden era. In his commentary on Genesis, the Reformer wrote that we, by comparison with the Patriarchs, are but "the offal of the world."[23] Such statements underline his pessimistic appraisal of the course of history, a presumption shared by Protestant Reformers in general.[24]

A crucial landmark in the history of the true church was the universal Flood that destroyed the original world and all of its human inhabitants, with the exception of eight individuals: Noah and his family. The Deluge ended the golden age of the long-lived Patriarchs, Adam's close descendants, who lived for several centuries.[25] The antediluvian world was a paradise compared to the world that succeeded it. Before the Deluge, the world and the faith of its inhabitants stood closer to the pristine perfection of creation. After the Flood, world history was characterized by degeneration, both of nature and morality.[26] The life-span of human beings was considerably reduced, procuring that man dies before he acquires wisdom. The longevity of the antediluvian Patriarchs was considered crucial to the preservation and transmission of knowledge.[27] By reputation and sanctity, the most remarkable men of the whole of humankind lived in the antediluvian world.[28] The association of the antediluvian men with wisdom was widely shared, and can be seen as attributable to their several-centuries' longevity. Even so, the Patriarchs perished in the Flood because they conceived too high an opinion of their own wisdom and other gifts.[29] The Deluge took place precisely because Adam and Eve's descendants—with the exception of Noah—no longer obeyed the Word and had lapsed into idolatry and lechery. The Patriarchs were guilty of sins against the first tablet of the Decalogue. What happened to them would later befall empires: man is his own enemy, Luther argued. One could thus lose the gift of the membership of the true church.[30]

23 Luther, "Enarratio," 42: 265.
24 Grafton, *What Was History*, 111.
25 Luther, "Enarratio," 42: 264 (Commentary on Gen.6:1).
26 Barnes, *Prophecy*, 39–40.
27 Harrison, *Fall of Man*, 164, 167; Luther, "Enarratio," 42: 270 (Commentary on Gen.6:1–2).
28 Luther, "Enarratio," 42: 265 (Commentary on Gen.6:1).
29 Ibid., 264 (Commentary on Gen.6:1).
30 Ibid., 265 (Commentary on Gen.6:1).

Luther used the Flood as analogous to his conflict with the Catholic Church: God let the pope and his entourage believe that they were the church; but they had, in fact, excluded themselves from the true church and were rejected because they no longer lived by His Word. For Luther, the outsider Noah represented the true church. His announcement of the impending Flood was met with incredulity, his prediction was said to be impious and heretical. The small number of survivors after the Flood indicated that the size of the community of true believers was of no importance. The divine destruction of humankind had proven God's care for the church: He preserved and governed the church, regardless of whether human beings understood His ways.[31] Luther designated himself as a new Noah who "was the true Church in his generation," remaining steadfast and worshipping God truly.[32] That the world had never been in such a bad condition was viewed by Luther as an imperative to persist in faith.[33]

The Wittenberg Reformer linked the imagery of Noah and the Flood to the Apocalypse that he believed was imminent. Protestant Reformers in general regarded their own age as decisive in the cosmic struggle between God and the devil.[34] The intensely apocalyptical nature of early Lutheran thought was stimulated not least by Luther's own prophecies of the Last Day.[35] Swiss Reformer Huldrych Zwingli (1484–1531) named his German counterpart "a contemporary Elijah," a prophet who preached God's Word as the end of times approached.[36] When lecturing about God's promise to Noah that the earth will one day be restored, Luther reminded his audience to bear this text in mind when they came across signs announcing the Last Day.[37] Comets, storms, and winds were perceived as unusual natural signs signifying that the end of time was near, reminding their observers of the natural upheavals that were described to have accompanied the universal Flood.[38] Although present since the fifteenth century, apocalyptic thinking in Lutheran areas became "more explicit, more reasoned, and more comprehensive" throughout the sixteenth century.[39] This is testified by the high number of popular pamphlets on the theme which circulated through Lutheran centers in Germany, even increasing in number after

31 Ibid., 332–333 (Commentary on Gen.7:16–24).
32 Maxfield, *Luther's Lectures*, 166–168.
33 Oberman, *Luther*, 77.
34 Barnes, *Prophecy*, 44, 139.
35 Kusukawa, *Transformation of Natural Philosophy*, 125.
36 Kolb, "Luther's Function," 211.
37 Luther, "Enarratio," 42: 353 (Commentary on Gen.8:22).
38 Barnes, *Prophecy*, 92.
39 Ibid., 3–5.

Luther's death in 1546.[40] Despite the fact that some authoritative astrologers deferred the end of the world by another 500 years, Luther believed it to be imminent: the Roman Catholic church was foundering in the mire of the errors of the papacy and fanatics. He believed as a consequence that the advent of the last day had been hastened.[41] Citing Matthew 24:22 ("For then shall be great tribulation, such as was not since the beginning of the world to this time, no, nor ever shall be"), he observed that Christ had predicted that the days would be shortened for the sake of his elect.[42]

Textbook prefaces and declamations on (natural) philosophy and the liberal arts circulating at Wittenberg University in the first half of the sixteenth century clearly echoed Luther's understanding of the true church, the Reformer's thoughts on the golden age of the Patriarchs and on the Deluge, and the utmost importance of the transmission of the divine promise, paired with the importance of *translatio studiorum*. *Translatio studiorum* is not used in the usual sense of a transmission of written texts from one geographical, political and linguistic context to another within the horizon and the limits of the formation of European culture, as has been the manner in which it has been deployed in other works, but as a conceptual transmission of knowledge and disciplines.[43]

2 The True Church in World History

The image of Wittenberg University as the true church was cultivated through the highly authoritative *Chronicon Carionis*, the preeminent textbook on world history used at Reformed universities in the Holy Roman Empire. Co-editors Philipp Melanchthon and Caspar Peucer depicted Wittenberg University as one in a historical series of schools that warranted description as sites for the teaching of God's Word.[44] The Reformed interest in history came about, among other things, from one of the counterarguments with which they were often confronted, eloquently encapsulated by Charlotte Methuen in the following question: "Where was your Church fifty or a thousand years ago?"[45] Novelty, after all, was considered the essence of heresy in early modern times.[46]

40 Ibid., 60.
41 Luther, "Enarratio," 42: 270–271 (Commentary on Gen.6:2).
42 Ibid.
43 A likeminded approach is used in Tullio, "Translatio Studiorum," 1.
44 Fraenkel, *Testimonia Patrum*, 133.
45 Methuen, "History and Heresy," 5.
46 Hamm, *Reformation of Faith*, 255.

The *Chronicon Carionis* had a complex genesis.[47] The manuscript was originally written by Berlin court astronomer and former fellow student of Melanchthon, Johannes Carion (1499–1537). He had sent the book to Melanchthon to have it published in Wittenberg. In 1532, after Melanchthon had reworked the manuscript, the *Chronica durch Magistru Johan Carion, vlessig zusammengezogen, meniglich nützlich zu lesen* was published by printer Georg Rhau. In 1558, a second edition of the chronicle was published in Latin. For the *Chronicon Carionis latine expositum et auctum multis et veteribus et recentibus historiis, in narrationibus rerum Graecarum Germanicarum et ecclesiasticarum*, Melanchthon and Caspar Peucer had reworked the content of the book together. A third, thoroughly revised edition was published by Peucer in 1572. (It is this edition that we use for the present study.) Between 1532 and 1572 the general tenor of the book had become strongly anti-papal.[48] The importance of the *Chronicon Carionis* can hardly be underestimated. It came into use in the sixteenth century as the standard textbook for universal history at Reformed universities in the Holy Roman Empire, being used by professors who had been appointed to these first chairs of history thanks to the efforts of Melanchthon and who, as students, had been taught history by Melanchthon and Peucer. Melanchthon was one of the first advocates for adding history to the curriculum and one of the first university professors in history to lecture on the basis of a printed textbook. Peucer succeeded to this chair after Melanchthon's death in 1560.[49] Throughout the sixteenth century, the chronicle was made accessible through translations in both Latin and vernacular, ranging from Low German to Czech, Italian, Dutch, and Icelandic.[50]

In the *Chronicon*, the true church was created by God with the creation of Adam, along with the gifts of God to humankind to be appropriated to glorify God, a church that was to obey God.[51] World history or universal history, from a Lutheran perspective, was in fact a history of the true church, the course of events unfolding in accordance with God's providential rule.[52] The authors ranked the creation of the church second among the most important things God wanted man to know, the first being the creation of the world and of man. Throughout history, the true church had found different refuges, such

47 Gemeinhardt, "Das Chronicon Carionis," 115–125.

48 For a fundamental study of the *Chronicon Carionis* and of Reformed history in general, see: Lotito, *Reformation of Historical Thought.*

49 Ibid., 12.

50 Ibid., 218.

51 *Chronicon Carionis*, 12, 13.

52 Ben-Tov, *Lutheran Humanists*, 39; Barnes, *Prophecy*, 140.

as Wittenberg University, and in so doing had ensured the transmission of true doctrine. The transmission of true doctrine—for Melanchthon, a learned faith—was key in his conception of the history of the church. Melanchthon depicted members of the primitive church as leading a perfect life, an example he wanted to be emulated by the contemporary descendants of the true church.[53] Peucer and Melanchthon emphasized that illustrious testimonies in all times proved that the passing down of astronomy or the doctrine of the measurements of the year, together with knowledge about creation, the Fall, and the church, from the first parents to posterity, were not fables but true.[54] They foreshadowed, moreover, the dual identity of Wittenberg University. Melanchthon pursued the imagery of the academy beyond the temporal dimension of this world; according to the preface to his commentary on Aristotle's *Liber de anima* (1553), there were two places of learning: the temporal settings of our earthly experience, and an eternal academy that awaits man after his mortal life where he acquires a perfect insight into nature and the causes of things, shown to him by God Himself.[55] The idea of the university as the true church surfaced at different occasions in Wittenberg academic life. In his 1542 oration on astronomy and geography, Rheticus, then dean of the Arts Faculty, explicitly identified the scholarly meetings at Wittenberg as assemblies of the church.

> And because these our assemblies are part of the Church, every time we gather together, every time we think of our studies, let pious prayers be added, and let us pray to God, the Father of our Savior Jesus Christ, that He may protect the Church and honest studies, and save and guide us.[56]

---

53 Fraenkel, *Testimonia Patrum*, 139.

54 *Chronicon Carionis*, 13–14.

55 Melanchthon, "Baumgartnero, filio," 7: 1124. The preface was addressed to the son of Hieronymus Baumgartner (1498–1565), a senator of Nuremberg, to whom Melanchthon addressed the preface of his previous commentary on *De anima* that was published in 1540; see: Melanchthon, "Hier. Baumgartnero," 3: 907–914.

56 "On Astronomy and Geography," 113; Rheticus, "Oratio de astronomia et geographia," A3^{v}–A4^{r}: "Et quoniam hi nostri coetus pars sunt Ecclesiae, quoties convenimus, quoties de studiis cogitamus, accedent pia vota, et oremus Deum, patrem liberatoris nostri Iesu Christi, ut Ecclesiam et honesta studia defendat, nos servet et regat." Several declamations of the Wittenberg scholars Rheticus and Reinhold were mistakenly attributed to Philipp Melanchthon due to their insertion in the *Corpus Reformatorum*, compiled and edited by nineteenth-century scholar Karl G. Bretschneider. According to Timothy Wengert, the *Corpus Reformatorum* is one of the conduits for the transmission of the axiom according to which Melanchthon was seen as the *praeceptor Germaniae* and of his status as a tireless ghostwriter of declamations for fellow professors (Wengert, "Caspar Cruciger," 419). The

The dean reminded his audience that Christ had promised that "He would be there wherever two or three gathered in His name."[57] The scholarly and religious collective bodies clearly coincided. At the end of the 1540s, Erasmus Reinhold expressed an identical view in an official invitation to the upcoming proclamation of the master students. In his function as dean of the Arts Faculty, Reinhold invited doctors, students, and masters "both to witness this proclamation and to supplicate God to protect the churches, the study of doctrine, the refuge of the churches and education in these regions and in this town."[58]

---

tenth and eleventh editions contain no fewer than 180 declamations by Melanchthon, the accuracy of which has been challenged since the end of the nineteenth century. Enquiry into the original sixteenth-century prints and into the sixteenth-century compilations used by Bretschneider shows that there is not sufficient evidence to support the possibility of Melanchthon's authorship. In 1542 the Nuremberg printer Johannes Petreius had issued both orations in a single standalone edition, entitled *Orationes duae, prima de astronomia et geographia, altera de physica, habitae Wittebergae a Ioachimo Rhetico, professore mathematum*. The dedicatory letter was addressed to Heinrich Widnauer, the mayor of Rheticus's hometown Feldkirch and was written by Rheticus himself. In it he explains to the mayor that these types of public and solemn orations pertained to his function as dean. Melanchthon's name was mentioned only once, with reference to his encouraging Rheticus to take on mathematical courses. Bretschneider extracted the orations from the 1558 edition of *Liber selectarum declamationum Philippi Melanthonis*. Its dedicatory letter clearly states that the book contains contributions by Melanchthon and others. The original titles of the relevant declamations in the 1544 edition still read *Oratio de physica habita Wittebergae, a Ioachimo Rhetico, and Oratio de astronomia et geographia, habita Wittebergae, a Ioachimo Rhetico*. By the time of the 1558 edition—the edition used by Bretschneider—the attribution to Rheticus had vanished. In short, there is no evidence to support any claims for Melanchthon's authorship; see: Burmeister, *Georg Joachim Rhetikus*, 1: 32, 69; Green and Murphy, *Renaissance Rhetoric*, 301; Koehn, "Philipp Melanchthons Rede," 1289–1290; Kraai, "Rheticus' Heliocentric Providence," 19, 94–95; Kusukawa, "Introduction," xiii; *Liber selectarum declamationum Philippi Melanthonis* (1544), 2^{v}–3^{r}; *Liber selectarum declamationum Philippi Melanthonis* (1558); Melanchthon, "De astronomia et geographia," 11: 292–298; idem, "De physica," 11: 555–559; Müller, "Zur Chronologie," 119; Rheticus, "Oratio de astronomia et geographia," A2^{v}; Pettegree, *Book in the Renaissance*, 91.

57 Rheticus, "Oratio de astronomia et geographia," A4^{r}: "[…] ad hos congressus dictum Christi pertinent, cum ait: Se interfuturum, ubicunque duo aut tres in nomine ipius convenirent."

58 "Decanus collegii philosophici in Academia Witebergensi," B5^{r}: "[…] et ut testes sint huius renunciationis, et ut simul nobiscum precentur studia, ut Ecclesias, doctrinae studia, hospitia Ecclesiarum et disciplinam clementer in his regionibus et in hoc oppido tueantur." Whether this invitation was issued as a circular or as a public announcement remains unclear. Reinhold added a wish-list of potential attendees, but unfortunately this was not included in print. This is one of the two hitherto unnoticed official statements issued by Reinhold in his capacity as dean of the Arts Faculty in 1549, found in the Berlin State Library. The untitled collection in which the statement appears was published by Vitus Creutzer (Veit Kreutzer) in 1549. It compiles orations, questions, and what seem to be

The assembly of scholars present on that particular date, 20 August, was thus no different from a religious gathering. On other occasions besides the highlights of the academic year, teachers and students obviously met on a regular basis. Reinhold argued that these gatherings, like church congregations, were protected by God and depicted them as assemblies of the true church.

> It is becoming of virtuous minds to be inclined towards such labors, that we do not doubt to be the care of God, who has not created humankind in vain but always unifies the Church, He keeps even now the many gatherings of teachers and students unharmed.[59]

A later reference indicating the identification with the true church is found in a letter written by Peucer in 1573 to Landgrave of Hessen-Kassel William IV—dubbed the Wise—on the subject of a new star that had appeared in November 1572. Kassel was an important extra-academic astronomical center in the Holy Roman Empire. At the time, Peucer was still active as an astronomer but was no longer professor in higher mathematics at Wittenberg, having assumed the chair of medicine from 1559 onwards.[60] He styled the correspondence between him and the Landgrave, who himself was an active propagator of the Reformation,[61] as one between men "who are, and publicly declare to be, true members of the church of the son of God."[62] The notion of scholars as adherents of the true church was well absorbed into broader Lutheran learned culture.

3 Origins of Astronomy

Wittenberg astronomers Peucer, Reinhold and Rheticus embedded the history of their art in the universal history of the true church.[63] Such disciplinary his-

---

official statements made by the dean of the Arts Faculty, entitled "Decanus collegii philosophici" possibly supplemented with the name of the dean being referenced in each case.

59 "Decanus collegii philosophici Erasmus Rheinholt Salveldensis," B3^r: "Talibus laboribus favere bonas mentes decet, quos quidem Deo curae esse non dubitamus, qui cum non frustra condiderit genus humanum, sed semper inde colligat Ecclesiam, servat etiam coetus aliquos docentium et discentium [...]."

60 Brosseder, *Im Bann der Sterne*, 23; Wege, "Peucer," 7: 629.

61 Ribbed, "Wilhelm IV," 43: 32–39.

62 Peucer, letter to Wilhelm IV, Wittenberg, 1 January 1573, 3: 122: "[...] qui sumus et profitemur nos esse vera membra Ecclesiae filii Dei [...]."

63 Asaph Ben-Tov reached the same conclusion with respect to the transmission of Greek throughout history. Lutheran humanists paralleled the history of the transmission of

tories as featured the transmission of a certain art or discipline throughout history pertain to the Renaissance genre of prefatory disciplinary histories. The histories of astronomy were surveys of the past of astronomy, emerging in dedicatory letters, prefaces, prolegomena to astronomical treatises, and introductory lectures on the subject.[64] Unanimously, the astronomers identified Adam as the founder of the astronomical art and they also included several Hebrew Patriarchs as early astronomers.[65] In itself, this account of an Adamic origin was not unusual in disciplinary histories of this period: histories of mathematics setting forth the Adamic origin of an art are found throughout the entire early modern period in different confessional settings.[66] The ideal of original knowledge, a *prisca sapientia*, was also a common Renaissance theme. Although there was no fixed canon regarding the other names enumerated in these prefatory histories of astronomy, all authors under discussion agreed that astronomy originated with the first father of the human race.[67] The recurrent Adamic theme in these disciplinary histories of astronomy has been aptly explained as a motif to underline the praiseworthy character of the art, and as a strategy in defense of the legitimacy of astronomical and astrological practice.[68] One respect in which these Protestant disciplinary histories differ, however, is in the fact that they establish a link between the disciplinary history and divine providence.[69]

In 1551, Caspar Peucer inserted a "Succession of astrologers from the first father until our age" in the astronomy textbook *Elementa doctrinae de circulis coelestibus, et primo motu*, encompassing a historical sequence of astrologers running from Adam to Copernicus.[70] Despite the title's reference to a chronological index as a series of astrologers, Peucer explained in the accompanying dedicatory letter to the Duke of Saxony that he did not want to speak about mantic or divinatory practices in his astronomy textbook, but about the motions of the stars. His use of the word *astrologus* illustrates the way in which the terms astronomer, astrologer and mathematician were regarded, at

---

Greek within the universal history of the Church. The lineage of the Greeks was connected to the Old Testament by tracing their origins back to Noah's progeny, notably to his grandson Ion; see: Ben-Tov, *Lutheran Humanists*, 64, 212–223.

64 Jardine, *Birth of History and Philosophy of Science*, 261–262.

65 For example, Rheticus, "Oratio de astronomia et geographia," A5^v–[A6^v].

66 Goulding, "Histories of Science," 37; Popper, "Abraham," 106.

67 Jardine, *Birth of History and Philosophy of Science*, 261–262.

68 Goulding, "Histories of Science," 38; Howell, *God's Two Books*, 51; Popper, "Abraham," 106.

69 Špelda, "Search," 340.

70 Peucer, *Elementa*, A1^r–A5^v.

that time, as interchangeable.[71] The list comprises approximately 80 names, covering a period of 5,512 years (of the 6,000 years for which the world was believed to have endured), including Patriarchic, Greek, Arab, Latin, and German astrologers. Judging by the layout, Peucer distinguished five major periods in the history of astronomy, four of which are arranged under explicit headings: 1) before the Deluge (Adam to Noah), 2) after the Flood (Shem to Joseph), 3) a transitory phase without heading between the last Patriarch Joseph and the reign of Alexander the Great (Atlas to Callippus of Cyzicus, *b.*c. 370 BCE), 4) from Alexander until Christ's Nativity (Aristaeus, *fl.c.*350–330 BCE, to Sosigenes of Alexandria, fl. middle of the first century BCE)[72] and, finally, 5) from Christ's Nativity until Peucer's own day (Strabo, *b.*64/63 BCE-*d.*25 CE, to Copernicus). Peucer concluded the sequence with an enumeration of astronomers at or affiliated with Habsburg, ranging from Johannnes von Gmunden (c. 1382–1442), an astronomer active at the University of Vienna in the fifteenth century, to Copernicus.

Typical for the genre, few details are provided about astronomical accomplishments. The dates of birth are mentioned, often together with the dates of death, and are timed in relation to the eschatological events of creation, the Deluge, or Christ's Nativity. Peucer's index is in fact a concrete example of chronology applied to astronomical practitioners throughout history. Into a single chronological scheme, he harmonized facts about astronomers (or about those who discussed astronomers or astronomical practices) drawn from a variety of sources with astronomical data. Over a period of more than five millennia, astronomy is seen to have been passed down from father to son, from teacher to disciple, and from people to people. The index exemplified the humanist themes of *translatio studiorum* of astronomy, and the succession of astronomers that had become standard in sixteenth-century historiographical writing.[73]

The first 23 names, the *linea Christi*, pertain to the genealogical line from Adam to the Patriarch astronomers and eventually draws to a close with Joseph in the year 2310 after creation. Together they represent the Patriarchic astronomers living before and after the Flood. Peucer added the respective birth and death years of these men, attesting to his view of them as historical figures. Adam, for instance, was born 5,512 years ago and lived until the 930th year of the world; his son Seth was born when the world was 120 years old and died

---

71 Westman, *Copernican Question*, 31.

72 According to contemporary historiography, it is unclear whether Sosigenes was actually of Alexandrian origin; see: Huxley, "Sosigenes," 547.

73 Jardine, *Birth of History and Philosophy of Science*, 273–275.

when the world was 1,042 years old. The universal Deluge had taken place in the 1,656th year of the world when Noah was 600 years old. Besides being the earliest cultivators of the divine gift of astronomy, these Patriarchs were also the earliest members of the true church. One of Peucer's few annotations in the course of the chronology attests the simultaneous transmission of theological doctrine and scholarly skills by the Patriarchs.

> There is no doubt that when those excellent men taught their children about God, about the difference in kind between men and cattle, about the causes of human death [the Fall], about reconciliation with God, and about the assembly of the church, they pointed out the traces of divinity impressed upon human minds—namely, law and the knowledge of numbers—and showed many other witnesses of God in the order of the heavenly motions, and in the position and figures of the cosmic bodies.[74]

This symbiosis of what was later subdivided into theological doctrines and lay skills was something to which Peucer referred as "the wisdom of the golden age."[75] This notion underlines the earlier statement that these were not histories of theoretical progress: they were pervaded with the Lutheran sense of historical pessimism. Rather than the Fall, the Deluge was perceived as a turning point in the history of astronomical practice, closing the golden age of the Patriarchs. The primitive church as presented in the *Chronicon Carionis* was intended by Peucer and Melanchthon to serve as an example to posterity, inciting others to imitate the perfect lives of these men.[76] The historical Patriarchs thus pursued a pristine and worthy form of astronomy, commensurate with a golden age.[77] The example of the antediluvian astronomers served as models for contemplation and imitation so as to preserve the doctrine for the sake of the republic.[78] These earliest members of the true church were also the earliest practitioners of a learned faith of which the motions of the stars were an integral part.

74 Peucer, *Elementa*, X2$^{r–v}$: "Nec vero dubium est, illos summos viros cum docerent suam posteritatem de Deo, de discrimine generis humani et pecudum, de causis humanae mortis, de reconciliatione cum Deo, de collectione Ecclesiae, monstrasse vestigia divinitatis impressa mentibus hominum, noticias numerorum, et legem, et ostendisse alia multa testimonia de Deo in ordine motuum coelestium et positu et figura corporum mundi."

75 Ibid., 3^{r}.

76 Fraenkel, *Testimonia Patrum*, 139.

77 Granada, "Tycho Brahe, Caspar Peucer," 566; Howell, "Hermeneutics of Nature," 282.

78 Peucer, *Elementa*, X4$^{r–v}$.

The notion of a golden age in astronomy was a recurrent theme in histories of mathematics. Nicholas Popper depicts the average Reformed sixteenth-century history of mathematics as "a degeneration narrative, a problem of recovery instead of invention, of cleansing rather than expansion, with a Golden Age behind rather than ahead."[79] The transmission of astronomy continued, nevertheless, even after the Flood. When freed from incarceration in Egypt, the last Patriarch in the chronology, Joseph, sustained the traditional teachings by disseminating their contents into Egypt and the neighboring places of Palestine and Phoenicia.

> Joseph, when he ruled Egypt, [...] spread the teaching of the holy Patriarchs about God, about the emergence of the world, about the creation of the human race, about the fall and restitution of the first Parents, about the heavenly motions, and about nature; his teaching about the celestial motions, numbers and geometric demonstrations, and the whole history of natural things, have survived in Egypt and the neighboring areas of Palestine and Phoenicia until our own time.[80]

The Patriarchs symbolized an ideal of the church as a school, of a pristine symbiosis between erudition and faith. What appears as two different sets of knowledge had formed an organic whole in the golden age, aligning perfectly with Melanchthon's model of the true church as a school. Melanchthon's preface of a widely disseminated introductory textbook on natural philosophy, *Initia physicae doctrinae* (1549) provides a very apt illustration of this symbiosis in his depiction of Adam and Abel. Sitting in the grass under the shade of a large tree, immersed in the sweet smell of fruits, Adam instructed his son about heaven, the stars, the earth, the four elements, and the countries amongst other things.[81] This scene represented the beginning of the *translatio studiorum*, to be continued along the genealogical continuum of true teachers, from Adam to Abraham.[82]

---

79 Popper, "Abraham," 106.

80 Peucer, *Elementa*, A1^{v}: "[...] Ioseph [...] cum Ægyptum rexit, tum doctrinam sanctorum Patrum de Deo, de mundi ortu, de creatione generis humani, de lapsu primorum parentum et restitutione, de motibus coelestibus, de natura rerum, sparsit et mansit in Ægypto ac vicinis locis Palestinae et Phoeniciae, doctrina motuum coelestium, numerorum, demonstrationum geometricarum, et tota rerum naturalium historia ad nostra tempore usque."

81 Melanchthon, "M. Meienburgo," 7: 472–477.

82 Fraenkel, *Testimonia Patrum*, 66.

In Peucer's succession of astrologers, the Deluge marked a critical juncture in the historical transmission of the art of astronomy, "an eye of the needle through which human knowledge had to pass."[83] The names of the postdiluvian astronomers seem to be enumerated almost as a proof that the art had not been extinguished through time and retained some connection with its antediluvian past. Although the propagation of astronomy from the time of the sacred fathers until the beginning of the Olympiads (the ancient Greek calendar) and the means whereby this was achieved were shrouded in mystery, the ancient knowledge and the use of the art was proven by histories and poems. There is no doubt, wrote Peucer, that this study was preserved in Assyria, Chaldea, Phoenicia, and Egypt and was subsequently introduced into Greece.[84] The mathematician also included the names of the ancient Greek poets Homer and Hesiod, neither of whom was exactly well known for any astronomical efforts.

Rather than expounding a discourse of theoretical progress, the authors of the prefatory histories of astronomy seem to have been concerned with establishing a narrative of continuity. According to Erasmus Reinhold, God Himself wished for this continuity. In an oration on the life and work of the fifteenth-century astronomer Regiomontanus (1436–1476), the Wittenberg astronomer argued that God inspired men like Regiomontanus to become great men for the purpose of preservation and transmission:

> [...] without divine inspiration he could not have thoroughly grasped so many and such varied arts, learnt so many languages [...] if God had not kindled and assisted the fire of his mind and heart. In that way, in His great goodness, He is wont continually to encourage and help masters in every field in order to restore and protect the arts that are necessary for life.[85]

---

83 Nothaft, "Noah's Calendar," 193.

84 Peucer, *Elementa*, A2ᵛ.

85 "On Johannes Regiomontanus," 238: Reinhold, "Oratio de Johanne Regiomontano," A3ᵛ: "[...] neque tamen sine adflatu divino tot tamque varias artes penitus percipere, tot linguas discere [...] nisi illa mentis ac pectoris incendia Deus et fovisset, et adiuvisset, ut solet immensa bonitate ad restituendas et conservandas artes vitae necessarias in omni genere subinde excitare, et adiuvare artifices." The content of Reinhold's declamation on Johannes Regiomontanus was attributed to Melanchthon, based on its insertion into Bretschneider's *Corpus Reformatorum*. The nineteenth-century editor had based his decisions, among other compilations, on an unspecified sixteenth-century edition of the *Liber selectarum declamationum Philippi Melanthonis*. This compilation, that was issued several times, clearly stated that the book contained contributions by Melanchthon and others. The Regiomontanus oration was originally printed in 1549 by Wittenberg printer

God did not want his gifts to be cast away in disgust and contempt, but instead to be cherished, treated with respect. In this regard, Reinhold reminded his audience of the will, effort and assiduity of Regiomontanus, who neither neglected the body and intellect given to him by God, nor chose pleasure or money over study. The astronomer from Königsberg had ensured the preservation and transmission of astronomy by means of his writings, instruments, travels, correspondence, and not least by means of the establishment of a printing house in Nuremberg. Reinhold portrayed Regiomontanus as a man worthy of imitation, who contributed to the preservation of the art of astronomy in his time, and, in so doing, acknowledged and was worthy of God's gifts.

The notion of a pristine Patriarchic astronomy continued well into the sixteenth century. As part of his 1574 inaugural lecture at Copenhagen University, Tycho Brahe sketched a prefatory history of astronomy, commencing with Adam, in which the earliest practitioners of astronomy also took the shape of the representatives of a true model of astronomy. This account too builds upon a lingering historical pessimism.[86]

4 Supporting a Bible-Based Theology

In the *Chronicon Carionis*, Melanchthon and Peucer portrayed the liberal arts as divine gifts, given to humanity at its creation in order to worship God.[87] In a similar fashion, the other Wittenberg astronomers repeatedly emphasized that the mathematical disciplines were a divine gift to man. "Let us embrace, love and cherish God's gift," Rheticus exclaimed about the science of heavenly motions in 1542 at a scholarly gathering.[88] He thus reiterated the Renaissance motif of astronomy as a gift from God. "Erudition," Melanchthon wrote in the preface to his commentary on Aristotle's *De anima* (1540), "is such a great ornament for the church that, once it is lost, it is impossible for the most impenetrable darkness not to follow. Even though there are other, more outstanding qualities, such as the true practice of faith, scholarship nevertheless sheds great

Vitus Creutzer. Its title explicitly mentions that the text was recited by Erasmus Reinhold. Melanchthon's name is absent. All evidence that remains is the mere inclusion of the oration in the *Corpus Reformatorum*, which is not sufficient. The 1549 oration on Johannes Regiomontanus is therefore considered to be authored by Erasmus Reinhold; see: Melanchthon, "De Iohanne," 11: 817–826.

86 Brahe, "Oratio de disciplinis mathematicis," 1: 148–149.

87 *Chronicon Carionis*, 14.

88 Rheticus, "On Astronomy," 117; idem, "Oratio de astronomia et geographia," [A6^{v}]: "[...] Dei donum amplectamur, amemus, et excolamus."

light on the Gospel [...]."[89] This citation discloses a first use of the liberal arts in relation to the true church, in their capacity as auxiliary sciences for reading Scripture. In assisting Luther in his translations of the Old and New Testaments, Melanchthon's role had consisted in determining the precise factual correctness of the translation.[90] Both Luther and Melanchthon deemed that the seven liberal arts could act as appropriate handmaidens to theology on the precondition that their application did not alter the content of what was written in the Bible.[91] Just as Melanchthon had redefined Scholastic natural philosophy in the service of Reformed theology,[92] so too was the functionality of the mathematical disciplines defined as an auxiliary science to Biblical lecture.

A speech on mathematical geography by the then dean Rheticus in 1542 gave a concrete example of how the liberal arts could shed light on the Gospel: by means of their geographical knowledge, scholars could contemplate the places where the church had once wandered.

> What is there in the prophetic scriptures on how the Church wandered and what was done there, from where one could judge, if that skillful description of regions was lacking? Every day in our prayers, we need to contemplate the land where the Son of God dwelt and was made a sacrifice, and we need to think of the place where the heavenly voice first sounded.[93]

A declamation by Reinhold as dean of the Arts Faculty some seven years later echoed the same line of reasoning. He argued that knowledge of the arts and languages was an ornament for the Church, because, besides their usefulness for life, they could illuminate the prophetic and apostolic writings. Without

89 Melanchthon, "Preface to the Commentary on the Soul," 912: "Tanquam decus est Ecclesiae eruditio, ut, hac amissa, non possint sequi denissimae tenebrae. Etsi enim aliae sunt excellentiores dotes, ut exercitia vera pietatis, tamen magnum lumen Evangelio addunt literae [...]."

90 Frank, "Luther and Melanchthon," 391–392.

91 Green, "Melanchthon's Relation to Scholasticism," 276–277, 281.

92 Bellucci, *Science de la nature*, 17.

93 Rheticus, "On Astronomy," 117; idem, "Oratio de astronomia et geographia," [A6^v]–[A7^r]: "Quid in scripturis Propheticis, quomodo vagata sit Ecclesia, quae res ubi gestae sint, unde iudicari posset, si deesset haec artificiosa regionum descriptio. Quotidie in precibus intuenda est regio, in qua filius Dei versatus est, et victima factus, cogitandus est locus, in quo primum sonuit vox caelestis." Another such example in the field of mathematical geography is a brief treatise on Biblical toponyms that Melanchthon added to a textbook in geography, *De dimensione terrae et fontibus doctrinae longitudinis et latitudinis locorum*, written by Caspar Peucer and published in 1550.

knowledge of languages and diction, after all, one could neither read nor comprehend such writings. Linguistic ignorance had led to disgraceful mistakes and confusion in relation to God that had persisted throughout many centuries. Knowledge of languages and diction, by contrast, were an aid to the preservation of the doctrine, Reinhold argued, that sheds such valuable light on human life.[94]

The specific utility of mathematical astronomy for the church was elucidated in Genesis 1:14, a recurrent citation in early modern writings vindicating the pursuit of astronomy and astrology: "And God said, Let there be lights in the firmament of the heaven to divide the day from the night; and let them be for signs, and for seasons, and for days, and years." According to Luther's commentary on Genesis, the verse made explicit the divinely intended purpose of these lights.[95] By observing heaven, man would be able to mark the passing of time, distinguish seasons and years, and detect divine signs such as meteors, eclipses, or rainbows or, such as would announce divine wrath or mercy. Whereas Luther endorsed the pursuit of astrology only as an intellectual pastime, Melanchthon used the passage in Genesis for legitimating not only astronomical but also astrological practice.[96] Both Rheticus's 1536 inaugural lecture upon assuming the chair of arithmetic *In arithmeticen praefatio*[97] and his *Oratio de astronomia et geographia*[98] advanced the notion that it was

94 "Decanus collegii philosophici Erasmus Rheinholt Salveldensis," B3^{r-v}.

95 Luther, "Enarratio," 42: 32.

96 Brosseder, "Writing in the Wittenberg Sky," 559; Kusukawa, *Transformation of Natural Philosophy*, 127.

97 *In arithmeticen praefatio*, [A6^{r}]. In the existing scholarship, both Rheticus and Melanchthon are credited with the authorship of this inaugural lecture. Sachiko Kusukawa, Barbara Bauer, and Charlotte Methuen all attribute the actual writing of the declamation to Melanchthon, while Rheticus is supposed merely to have recited it. However, besides its inclusion in sixteenth- and nineteenth-century compilations, there is no substantial reason why Rheticus should not be the author, as has also been asserted by Karl Burmeister, Jesse Kraai, and Stefan Deschauer. The *editio princeps* of the inaugural lecture was published by Joseph Klug (1523–1552) in Wittenberg in 1536 as *In arithmeticen praefatio Georgii Ioachimi Rhetici*. From 1541 onwards, the inaugural lecture has not been reprinted independently. In 1582, the printer Martin Mylius inserted the arithmetic preface in the *Chronologia scriptorum Philippi Melanthoni* and from there it found its way into Bretschneider's *Corpus Reformatorum*. The mere inclusion in the *Corpus Reformatorum* is not sufficient evidence to sustain any argument for Melanchthon's authorship; see: Bauer, "Gott," 156; Burmeister, *Georg Joachim Rhetikus*, 2: 56; Deschauer, *Die Arithemtik-Vorlesung*, v; Kraai, "Rheticus' Heliocentric Providence," 19; Melanchthon, "Praef.," 11: 284–292; Methuen, "Role of the Heavens," 385–403, Mylius, *Chronologia scriptorum Philippi Melanchthonis*.

98 Rheticus, "Oratio de astronomia et geographia," A5^{v}–[A6^{v}].

God's wish that man would observe the course of the sun and the fixed boundaries in the year. Astronomy served a chronological purpose: to keep track of regular providence, the orderly course of time, and of history in its continuation. The discipline of chronology enabled the integration of all historically recorded events within a single scheme of universal history.[99]

> For if these masters had not preserved this sequence of seasons for us, what darkness would there be in religions and all of history, and what confusion of all life? We would not know which religions were earlier and which later; the succession of realms would be unknown and so would the beginnings and the propagation of the Church.[100]

A similar citation is found in Peucer's textbook *Elementa*:

> But God shows the certain beginning of the world, and then numbers the years, so that all of posterity would know at what time and by what intervals he would reveal himself; how he passed down his teaching; when the Son would desire to be seen among the human race; and where and when the Church would be established. And so we assert that the teaching of the measurements of the Year is necessary, and that the effort to learn this, which is necessary to the Church, is pleasing to God. For this reason, the effort of learning is more pleasant to the Scholars because they understand that it is pleasing to God, and the masters willingly support these studies because they see the necessity of safeguarding this ornament of the Church. Therefore, in all histories and descriptions of Empires, how great a light is offered by the progression of years! Which, because it is preserved in the Church, we acknowledge it to have been done by the particular counsel of God [Gen.1:14], and on account of this we give thanks to God for this favor and will not let this teaching be extinguished.[101]

---

99 Grafton, "Dating History," 74–85.

100 Rheticus, "On Astronomy," 117; idem, "Oratio de astronomia et geographia," [A6v]: "Nisi enim artifices hanc temporum seriem nobis conservassent, quae essent religionum atque omnium historiarum tenebrae, quae confusio totius vitae. Non animis retro legere mundi initia possemus, non sciremus quae religiones priores, quae posteriores fuissent. Esset ignota series imperiorum, nescirentur Ecclesiae initia et propagatio."

101 Peucer, *Elementa*, X5r–v: "Sed ostendit Deus certum exordium mundi, et deinde annos numerat, ut quo tempore, quibus intervallis sese patefecerit, quam doctrinam tradiderit, quando Filium conspici in genere humano voluerit, ubi, quibus temporibus Ecclesiam instaurarit, tota posteritas sciret. Fateamur ergo et necessariam esse doctrinam de Anni metis, et Deo gratum esse hunc discendi laborem Ecclesiae necessarium. Eamque ob

The celestial signs equally played an important role in providing a firm framework for historical time. Dated astronomical eclipses established chronological benchmarks to which historical events could be related.[102] In the dedication of Reinhold's commentary upon the *Theoricae novae planetarum* by Georg von Peurbach (1423–1461), the astronomer elaborated on the noting down of eclipses as an essential part of historiographical activity.[103] Many histories, Peucer wrote, integrate the eclipses that foretold the downfall of kings and destruction of cities. The ancient Greek historian Thucydides (c. 460-c. 395 BCE), for example, had written about a solar eclipse during the first year of the Peloponnesian War (431–404 BCE), an eclipse bringing about so great a darkness during the middle of the day that the stars in the heavens became visible. Reinhold indicated that, besides their chronological use, celestial signs were also considered messengers of divine wrath. That particular eclipse during the Peloponnesian War had announced a horrifying disease that combined a scorching fever with the rotting of the forearms in only a few days, or the mortification of the feet; the annihilation of cities, and the massacre of armies.[104]

The Wittenberg astronomers are thus seen to reinterpret timekeeping as an instrument to shed light on the history of the true church, and thus as a means to keep track of sacred history. Holy Writ contained a chronological framework of eschatological events regarding the first advent of Christ and the restoration of the church precisely so that later ages could situate themselves within the unfolding of sacred history. By counting years and days—the chronological purpose of astronomy—, man could relate himself to these events and could keep track, as Rheticus indicated, of the propagation of the church. Scripture was one of a wide variety of sources that the chronologist drew upon in the practice of chronology, ranging from the Bible and the books of ancient Roman and Greek historians to astronomical tables. Reminiscent of Katherine Park's remarks on timekeeping in monastic life during the early and central Middle Ages, observation here turns into observance, "focused on the transcenden-

causam et Scholasticis discendi labor iucundior sit, quia Deo gratum esse intelligunt, et principes libentius haec studia iuvent, quia vident se ornamentum Ecclesiae necessarium tueri. Deinde et in omnibus historiis, et Imperiorum descriptionibus quantum lumen est Annorum series. Quae quod in Ecclesia conservata, agnoscamus id singulari consilio Dei factum esse, et pro hoc beneficio Deo gratias agamus, et non extingui hanc doctrinam sinamus."

102 Grafton, "Dating History," 79.

103 *Theoricae novae planetarum*, A2^{r}–A8^{v}.

104 Ibid., A2^{v}.

tal meanings of terrestrial experience—while the pious observer, immersed in the divine order, became an instrument for measuring the orderly passage of time."[105]

5 Keeping it Pious

In his eulogy at Luther's funeral, Melanchthon credited the deceased for having brought to light the true doctrine and called back the church "from its pagan madness which pretends to invoke God, even though minds overwhelmed by Academic doubts flee God."[106] Melanchthon was of the opinion that, "many and great disputes are held in the Church, which cannot be expounded without erudition."[107] For example, someone who spoke about the different parts of the soul could never do this skillfully without being versed in the natural philosophy that was taught in the schools.[108] Even though Melanchthon had denied that the metaphysical aspects of natural philosophy were of value to theology, he still gave a role in theological debate to a general education in natural philosophy.[109] He was convinced that erudition prevented the members of the church from developing an erroneous or barbarous theology.[110] In a similar manner, Rheticus ruminated upon the topic of fate: "How will one who is inexperienced in natural philosophy speak about this, how will he overthrow the axioms of the Stoics or Peripatetics, which he does not know?"[111] In the same declamation, the astronomer insisted on an educated church:

> Therefore, natural philosophy is also to be embraced, if we do not want the Church to be ignorant [...]. Because they do not restrain their minds

105 Park, "Observation in the Margins," 22–23.

106 Melanchthon, "At Luther's Funeral," 258; idem, "In funere Lutheri," 11: 727: "[...] totam Ecclesiam ab ethnico furore revocavit, qui fingit Deum invocari, etiamsi mentes academica dubitatione oppressae fugiant Deum."

107 Melanchthon, "Preface to the Commentary on the Soul," 149; idem, "Hier. Baumgartnero," 3: 912: "Multae enim et magnae controversiae agitantur in Ecclesia, quae sine eruditione nequaquam explicari possunt."

108 Rheticus, "Oratio de physica," B2v.

109 Green, "Melanchthon's Relation to Scholasticism," 276–277. Given the scope of early modern natural philosophy, many topics were included in its discourse that, at a later stage, would be called metaphysical or theological, cf. Osler, "Mixing Metaphors," 92.

110 Fraenkel, *Testimonia Patrum*, 125–126.

111 Rheticus, "On Natural Philosophy," 137; idem, "Oratio de physica," B2v: "Hic quid dicet imperitus physicaes, quomodo labefactabit stoicorum aut peripateticorum sententias quas ignorat."

> with the fixed limits of the sciences, the ignorant stray easily from the path, and it is more difficult to correct them, because they neither know nor observe the laws which one has to follow in teaching like a pattern, and they do not have true hypotheses. If we want to avoid these disadvantages there certainly is great need for us to improve and refine our minds by true erudition. […] If the light of erudition became extinct in the Church, various kinds of madness would ensue, as has happened before.[112]

In this regard, a rational exploration of theological dogma was "a legitimate weapon in the controversialist's armor."[113] A major responsibility follows from these insights for the respective scholars, however: keeping (natural) philosophy pious in the light of theological debate. Rheticus unambiguously stated before his audience that there is no doubt that the church has need of an erudite and "undeteriorated" philosophy, whilst he argued that badly constructed opinions of natural philosophers had previously caused great upheavals in the church.[114] Piety in this regard means agreement with Holy Scripture or God's Word.

Melanchthon's exposition of ancient philosophical schools in his highly influential textbook on dialectic or the art of argumentation, the *Erotemata dialectices*, issued in 1547, shows that the chief criterion for the validity or trustworthiness of such a philosophy was its agreement with true doctrine.[115] When enlightened by God, man ought to avoid the false opinions of philosophers, or literally "cause-givers" (*aitiologikoi*). The *praeceptor Germaniae* explains that there is only one eternal and firm truth, even in the liberal arts, and men should refrain from inventions so that this truth would remain a light in life. Wavering in this regard meant *petulentia* or impertinence, and was perceived as pernicious to human life, and as insulting towards God who was considered to be

112 Rheticus, "On Natural Philosophy," 137–138; idem, "Oratio de physica," B2^{v}–B3^{r}: "Est igitur et physica doctrina complectanda, si volumus Ecclesiam non esse ineruditam […]. Indocti quia ingenia non coercent metis doctrinarum, facilius deerrant à via, et difficilius sanari possunt, quia nec norunt nec observant leges, quas in docendo sequi tanquam normam, necesse est, deinde hypotheses veras non tenent. Haec incommoda si vitare volumus, profecto excoli et expoliri ingenia vera eruditione valde opus est. […] Si igitur lumen eruditionis in Ecclesia extinctum fuerit, deliria exorirentur varia, ut antra accedit."

113 Brockliss, "Curricula," 2: 597.

114 Rheticus, "On Natural Philosophy," 136; idem, "Oratio de physica," B2^{r}: "[…] tamen nihil dubium est, Ecclesiae erudita et sincera physica opus esse."

115 Melanchthon, "Erotemata dialectices," 13: 655–658. On the different editions of Melanchthon's handbook in dialectic and the transformation of the discipline thereby brought about, see: Burnett, "Educational Roots," 299–317.

the source of truth. According to Michel-Pierre Lerner, Melanchthon's use of the term "*petulentia*" presents a moral and intellectual critique, denouncing those who adhere to absurd opinions and who delight in the perverse nature of monstrous opinions.[116] Scripture thus emerges as a touchstone for (natural) philosophy. When comparing Stoic, Academic, Epicurean, and Aristotelian philosophies, God's Word is essential:

> When we read what repulsive opinions about God they spread with dreadful arrogance, we shudder to think that the devil overthrows minds not governed by the divine light, and we oppose to them the testimonies by which God has revealed Himself and fortify our hearts with true opinions not only against the Epicurean madness, but also against Stoic gibberish and Academic doubt, and we desire that the true knowledge of God is kindled and confirmed in us.[117]

Epicurus, Democritus, and the like, showed what excrescences human reason was capable of when deprived of the light of Scripture. According to Jeanne Fahnestock, Melanchthon's exposition on the sects came with a warning, notably against disdaining accepted opinions out of a love of novelty and controversy, something of which the Protestants themselves were often accused.[118] Holy Writ thus enabled the natural philosopher to discern falsehoods in pagan philosophies, "falsehoods" being defined as divergences from true doctrine. The importance of the *Erotemata dialectices*—of which 47 editions were published by 1603—cannot be understated. The third edition of 1547—which is used here—was widely used in the schools and universities of Northern Europe.[119]

Melanchthon's appraisal of the teachings of Aristotle underwent something of an evolution. In his Arts Faculty curriculum reform of 1518, the then rector nearly eliminated all lectures on Aristotle, mainly motivated by his dislike of the central role the Stagirite played in Scholastic theology.[120] In 1536, however,

116 Lerner, "Aux origines de la polémique anticopernicienne (II)," 435.

117 Melanchthon, "Erotemata dialectices," 13: 658: "Cum vero legimus, quam tetras opiniones horribili audacia de Deo sparserint, cohorrescamus cogitantes, quo diabolus impellat mentes, quae non reguntur luce divina, et opponamus testimonia, in quibus Deus se patefecit, et veris sententiis pectora muniamus, non solum contra Epicureos furores, sed etiam contra stoica deliria et academicam εποχην seu dubitationem, et accendi et confirmari in nobis veram agnitionem Dei petamus."

118 Fahnestock, "Introduction," 54.

119 Ibid., 3.

120 Wengert, "Philip Melanchthon," 17–18.

another curriculum reform saw an extensive restoration of Aristotle's teachings, motivated by the need for a reliable philosophical foundation for the Reformation, which led to the development of a Protestant Scholasticism.[121] As at other universities, Aristotle was at the center of the academic program.[122] Melanchthon's discussion of the Aristotelian, Epicurean, Stoic, and Academic schools was revealing of the respective discords between these philosophies, on the one hand, and the true doctrine, on the other. Of the four schools enumerated, the one to be preferred is that which "teaches true things for the most part, establishes the arts from true principles, and contains a minimum of errors."[123] In Melanchthon's opinion, the Aristotelian school is the one that best met these criteria. Melanchthon praised Aristotle for having divided philosophy correctly into three disciplines: dialectic, physics, and ethics. However, the Stagirite erred about the eternity of the world. On this subject, Melanchthon wrote, one should follow the teaching handed down in God's illustrious testimonies. Epicurean philosophy, full of "horrible ravings," was rejected due to the assumptions that the world is constructed out of atoms and governed by coincidence. From these premises, the Epicureans removed the final and efficient cause of the universe and denied the existence of God and His providential care. Stoic doctrine was dismissed among other things for its restriction of God's power to secondary causes. Finally, Melanchthon rebutted Academic doctrine because it professed that all things were uncertain, a position opposed to the notion of God's granting to humankind the gifts of arithmetic, geometry, philosophy, and ethics. God Himself ensured that the certainty of these doctrines was firm and unwavering. Melanchthon continued by arguing that the teachings of Democritus, the Epicureans, the Stoics, the Pyrrhonists, and others, distorted the simple doctrine and introduced monstrous views that corrupted the human mind and derided God. These opinions were espoused by minds that were not ruled by the divine light; such opinions were the work of the devil. In the *Erotemata dialectics*, Melanchthon wrote that demonstration prevailed over authority in matters of philosophy. This remained the case only in the absence of divine authority, however: if the Bible taught clearly about a particular matter, divine authority prevailed over both demonstration and authority. Demonstration he defined as the testimony of an extraordi-

121 Kirschner and Kühne, "Decline," 35.

122 Bellucci, *Science de la nature*, 145.

123 Melanchthon, "Erotemata dialectices," 13: 655: "Quare in delectu sectarum eam praeferamus, qua magna ex parte res veras tradit, et artes ex veris principiis extruit, et minimum habet errorum."

nary master of art.[124] Almost by definition, this *artifex* was an ancient natural philosopher whose work had become canonized and represented a tradition in the study of nature, because Melanchthon considered that "age, novelty, and inexperience complicated judgment."[125] He deemed that it was safest to study nature by means of the writings of ancient philosophers, and of Aristotle in particular, whose texts in his opinion showed very little deviation from Scripture. The prevailing opinion at Wittenberg held that a university was neither a place for novelty nor for philosophical debate for its own sake; learning was about serving religion and a Bible-based theology, not sowing doubt. The priority in education was to disseminate commonly accepted views; the priority when dealing with natural philosophy was compatibility with God's Word.[126]

The responsibility of learned scholars towards the true church was to prevent true doctrine from being perverted, and consisted of a continuous effort to (re)orient the human mind towards the Word of God. Like Luther, Melanchthon believed that the community of the true church was at risk of being corrupted.[127] Still, the *praeceptor* found it useful to teach these doctrines so that his students would not cleave to opinions for the sake of novelty and would not disdain received opinions. The stipulation of compatibility with holy doctrine was equally expressed by Rheticus who stressed that it was "to be led by a heavenly voice, in which God reveals more clearly both His nature and His will."[128] No less than Melanchthon, Rheticus stressed the necessity of a natural philosophy that was in concurrence with God's Word, and that was, as such, pious. Erudition was the ornament of the church as long as the philosophy in question retained its relation to God's Word. For Melanchthon, the submission of the mind to true doctrine also served a salutary purpose. In his preface to the 1540 edition of Aristotle's *De anima*, he emphasized that, having attained heaven, man will be glad to have defended in his life what is true and pleasing to God and to have renounced the ravings of the Epicureans and Stoics.[129] Hence, a pious natural philosophy was not only beneficial for keeping the human mind oriented towards God's Word, but it also prepared the human soul for eternal academy.

124 Ibid., 13: 707–709.

125 Ibid., 13: 707: "Impediunt iudicium, aetas, novitas, et imperitia."

126 Bellucci, *Science de la nature*, 146, 149.

127 Ben-Tov, *Lutheran Humanists*, 213.

128 Rheticus, "On Natural Philosophy," 136; idem, "Oratio de physica," B2^r: "Sed assentior regendas esse has cogitationes voce caelesti, in qua Deus magis perspicue, et naturam suam et voluntatem patefecit."

129 Melanchthon, "Hier. Baumgartnero," 3: 914.

6 Scholar-Protectors

As befits a true church, the university of Wittenberg had to ensure the transmission and thus survival of the learned faith, of the liberal arts, and the true doctrine. In the words of Erasmus Reinhold, the learned acted as "custodians" of the doctrine about God, the Church, and the human race, for the sake of the church and the republic.[130] The urgency to make explicit the responsibilities of the learned as scholar-protectors of the arts, as documented in several texts by Rheticus and, especially, by Reinhold in the 1540s and 1550s, was fueled by the contemporary military-political situation. The territories of the Holy Roman Empire were riven by a threefold war during the first half of the sixteenth century: the domestic wars of religion between the Protestant princes and Emperor Charles v (1500–1558); the rivalry between the latter and the French King Francis i (1494–1547), and wars against the advancing Ottoman armies.[131] Baleful as these wars were in and of themselves, for the scholars of the Wittenberg Arts Faculty, they were especially abhorrent given how inimical they were to the field of erudition and how detrimental to the true church. Erasmus Reinhold looked upon the calamities of his time with grief and despair: "What era," he asked his readers, "is more hostile to philosophy than this one itself, in which the studies of these our arts are impeded through both domestic controversies and foreign wars?"[132] In 1542, Rheticus argued that the study of the arts required a peaceful homeland: "Let us therefore entreat God for everlasting peace, for the sake of the Church that is otherwise quite badly affected, and is retaining only some small remnants of erudition."[133] By 1553, however, the situation still had not improved. At the outset of the dedication of his posthumously published *Tabulae directionum*, Reinhold declaimed that the Holy Roman Empire formed the battleground upon which the Emperor fought the French King; internally, the empire was in the grip of domestic upheavals as well such as the war in Saxony and a recent uprising in the duchy of Brunswick.[134] And yet the Saalfeld astronomer was able to find some solace. Based on Scriptural inferences, he advanced the idea that the course of events was in accordance

130 "Decanus collegii philosophici Erasmus Rheinholt Salveldensis," B5^{r}.

131 Soen, "From the interim of Augsburg," 1: 549–564.

132 *Theoricae novae planetarum*, A4^{r}–^{v}: "Quae fuit aetas magis inimicae philosophiae, quam haec ipsa, in qua et domesticis dissensionibus et externis bellis impediuntur harum nostrarum artium studia."

133 Rheticus, "On Natural Philosophy," 133; idem, "Oratio de physica," B2^{r}: "Hanc [pacem] igitur à Deo perpetuam petamus, propter Ecclesias alioqui satis male adfecta, et ad has quantulasunque reliquias literarum retinendas [...]."

134 Reinhold, *Primus liber Tabulae directionum*, α2^{r}-α3^{v}.

with God's will. Empires had fallen as a result of His divine wrath over the crimes of the human race and those kings who had forsaken their divine duty as guardians of justice.[135] The idea central to the Lutheran confession—that of the divine promise of salvation for the members of the true church—was a notion that also underpinned the perception of the divine protection of the arts. Reinhold found hope in the divinely promised preservation of the church. In spite of the ruin of his native country and the ubiquitous chaos, he remained confident that God would save the true church:

> [It is] the will of God, who punishes crimes even among the ruins of the world; however, by his immense goodness, he preserves the flock [i.e., the Church], the custodian of the true doctrine, so that a certain part of humankind recognizes and glorifies the creator. Although I am brought to such considerations by grief when looking upon the conflagration of the country, it is not thereby alien to this place, because it also brings about a consolation about the preservation of the Church. We cling to this anchor, that it is entirely certain that the Son of God, just as He kept the Ark unharmed during the Flood, in it being the family of Noah both witnesses and custodians of doctrine, so now also His Argonauts will be saved.[136]

Imagery of the Flood and of the Argonauts evoked the idea of a small group of heroic, courageous men who strove for survival amidst chaos and adversity; just like the terrible state of the world that prompted Luther to persist in faith.[137] Just as Noah and his family were saved from perishing in the Deluge, the image of the anchor in the above citation underlines Reinhold's confidence that God would protect his Argonauts again. In Greek myth the crew of the ship named Argos set out on a quest for the Golden Fleece under the leadership of Jason.[138] Asaph Ben-Tov has highlighted the fact that the adventures of the Argonauts, for Philipp Melanchthon, were real past events preceding the Trojan war: their

---

135 Ibid., α2^{v}.

136 Ibid., α3^{r}: "[…] voluntatem Dei, qui et punit scelera et inter mundi ruinas: tamen immensa bonitate coetum servat, custodem verae doctrinae, ut pars aliqua generis humani conditorem agnoscat et celebret. Etsi autem dolor me intuentem patriae incendium, ad hanc commemorationem deduxit, tamen eo non est aliena ab hoc loco, quia simul consolationem complectitur de Ecclesiae conservatione. Hanc teneamus ancoram, quod certissimum est Filium Dei, ut in diluvio arcam servavit, et in ea familiam Nohae et doctrinae testes et custodies, ita nunc quoque argonautas suos servaturum esse."

137 Oberman, *Luther*, 77.

138 Schmitz, "Argonautae," 1: 280.

voyage underlined their courageous nature.[139] It is equally possible that Reinhold too thought of the Argonauts not as mythological figures but as historical personae. Besides the evident imagery of the ship, the image of the Argonauts proved to be a recurrent motif in the scholarship. In 1551, Caspar Peucer initiated his dedication letter to the duke of Saxony in his *Elementa* with a tale about one of Jason's Argonauts who, when he and his fellows were fleeing a hostile fleet, came up with an alternative escape route based on his knowledge of geography and astronomy.[140]

6.1 *The Turks*

When faced with the Turkish armies, the practice of the liberal arts was not merely under threat; the perception was that their continued existence was in the balance, fearing for the arts being entirely wiped out. From the 1520s onwards, the troops of Sultan Suleiman I the Magnificent (1494/95–1566) were advancing quickly in the Balkan, Eastern Europe, and along the southern shores of the Mediterranean. Various Turkish military campaigns in Hungary and Eastern Europe had brought Ottoman troops very close to the border of the Holy Roman Empire.[141] The ubiquitous European fear of the advancing Ottoman armies informed Reinhold's warnings of a potential end to the unbroken lineage of the *translatio studiorum*. The risk of obliteration of the liberal arts would cut the true church off from its roots with the Patriarchs.

> Long-lived men in the first ages of the world, Adam, Noah, Shem and Joseph, passed on this knowledge to coming generations, and its possession was preserved in Egypt from the time of Joseph to the age of Ptolemy, and somewhat longer, for over two thousand years. For such a long time God bestowed upon this knowledge a stable and fixed abode, among the

139 Ben-Tov, *Lutheran Humanists*, 53, 55. Melanchthon's view on the timing is in accordance with the Greek tradition that situated the heroic quest of the Argonauts one generation before the Trojan War; see: Schmitz, "Argonautae," 1: 280.

140 Peucer, *Elementa*, X2^{r}–^{v}.

141 Shaw, *History*, 87–105. In 1541, a part of Hungary became annexed to the Ottoman empire (Ottoman Hungary). For the remaining parts of the Hungarian Kingdom, King Ferdinand (1503–1564), the younger brother and future successor of Holy Roman Emperor Charles V, was expected to pay tribute to the sultan if he wished to be allowed to remain on the throne. Instead Ferdinand assembled an army from various European countries and Christian denominations in order to launch a new crusade against the sultan who successfully routed the attack in November 1542. Although a truce was signed in 1545, the war between the Habsburgs and Suleiman was resumed six years later, this time in the principality of Transylvania where it continued for another decade.

changes of the various monarchies. But then the wisdom of the principal monarchs was such that they knew that they ought to wage war against the enemies and not against the arts and in the midst of the wars they protected the abodes of erudition and the assemblies of the learned. Now, oh grief, that knowledge roams without protectors and without a fixed abode, so that one must fear the ruin of these arts. But let us pray to God the maker not to allow His gifts to perish.[142]

A similar wording is found in the dedication of Reinhold's commentary on Peurbach's *Theoricae novae planetarum*, issued in 1542. In the course of history, the collapse of various empires—Egyptian, Persian, Greek, Roman—had never yet hindered the preservation or protection of the arts. The Turks, however, represented an unprecedented threat owing to the manner in which they waged war: manifesting godlessness, cruelty, and lawlessness, they simply destroyed everything in their path.[143] Rheticus too voiced these concerns. In a letter to his students in October 1541, he wrote that the Turkish wars threatened the extinction of the arts in Egypt, Asia, and Greece.[144] In his oration on astronomy and geography, the mathematician expressed his concerns for learning and for the church in the face of the devastation and ruin wrought by the savagery of the Turks.[145] This apocalyptic fear of Turkish destruction of Western intellectual heritage had become a well-known trope in the decades following the Ottoman seizure of Constantinople in 1453,[146] and intensified from the 1520s onwards

142 "On Johannes Regiomontanus," 239; Reinhold, "Oratio de Johanne Regiomontano," A4^{v}: "[...] sic longaevi illi primis mundi temporibus, Adam, Noha, Sem, Ioseph posteris hanc doctrinam tradiderunt, ac possessio eius in Aegypto retenta est inde usque; à Ioseph ad Ptolemaei aetatem, et aliquanto amplius, plus duobus millibus annorum. Tamdiu stabilem et certam huic doctrinae sedem Deus attribuit inter varias Monarchiarum mutationes. Sed ea tunc fuit praecipuorum monarcharum sapientia, ut non cum artibus, sed cum hostibus bella gerenda esse scirent, et in mediis bellis literarum domicilia, et doctorum coetus protegerent. Nunc proh dolor vagatur haec sapientia sine patronis, et sine certa sede, ut interitus harum artium metuendus sit. Sed Deum autorem oremus, ne sua munera perire sinat."

143 *Theoricae novae planetarum*, A5^{r}.

144 Rheticus, letter to the students of the Wittenberg Arts Faculty, Wittenberg, October 1541, 3:43.

145 Rheticus, "Oratio de astronomia et geographia," A3^{v}: "[...] inter medios hostes [...] si ita ferent fata [...]."

146 For instance, Cardinal Bessarion (1403–1472), then Latin patriarch of Constantinople in exile, was largely driven by this fear when compiling a large library of Greek manuscripts out of concern for the preservation of Greek learning; see: Shank, "Classical Scientific Tradition," 127–128.

because of the military campaigns of Suleiman I in Eastern Europe.[147] Martin Luther echoed these concerns. David Grafton found no fewer than 5,000 references to the Turks in the German Reformer's literary output.[148] Even in conversation at home, as testified by the *Tischreden*, the Ottomans proved to be a recurrent topic.[149] For him, Satan's ultimate manifestations were the threat to the true church from within represented by the papacy, and the external threat represented by the military advancements of the Turks.[150] Both were a tools of God's retribution.[151] Especially during the early years of the Reformation, Luther's understanding of the course of history was determined by the prophecies of the Biblical prophet Daniel.[152] In 1530, he published a translation of the Book of Daniel because he believed that certain fragments applied to his own time and interpreted them as references precisely to the reign of the Turks and to the pope as antichrist.[153] Daniel's Last Day prophecy from the seventh chapter centered on the image of a woman riding a beast with seven heads and ten horns. The seven heads represented seven mountains and seven kings, five of whom no longer reigned, one who currently reigned and another who would reign in the future. The ten horns symbolized ten kings who were to reign over a kingdom along with the beast. The littlest horn was identified by Luther as the Turks. Daniel foretold that the little horn would pluck out three of the ten horns, meaning that it would subdue three kingdoms. Appropriated for a sixteenth-century context, Luther identified the little horn as the Turks who had conquered the Kingdoms of Egypt, Greece, and Asia Minor. He believed the Turks to be the wrath of God. Their fall would be a sign of the imminence of the Last Day.[154] Luther and his followers thus believed themselves to be in the chaotic end time preceding the final triumph of good over evil, in which the true church suffered from final persecution.[155] They believed that the enduring battle between God and Satan would culminate in the final triumph of good over evil.[156]

Like Luther, the Wittenberg astronomers identified their own era as chaotic, symbolized through their use of imagery connected to Moses's record of the

147 Guthmüller and Kühlmann, *Europa*, 668.
148 Grafton, "Martin Luther's Sources," 665–683.
149 Brecht, "Luther und die Türken," 26.
150 Barnes, *Prophecy*, 42.
151 Brecht, "Luther und die Türken," 9.
152 Barnes, *Prophecy*, 39–40.
153 Gritsch, "Luther as Bible translator," 65.
154 Barnes, *Prophecy*, 40.
155 Ibid., 44, 139.
156 Ibid., 19.

Flood. In spite of the arts being on the verge of extinction, Reinhold expressed his confidence that the church and the arts would survive based on God's promise to protect the church.

> But because God has promised that the Church would be His care, the Turks are not capable of razing it to the ground; what is left will be an abode of heavenly doctrine and of the other virtuous arts; there will be ordinary Rulers who foster the remnants of the arts. That is why our spirits are not yet broken, why we do not yet despair for the Church or our studies.[157]

The Turks would neither wipe out scholarship nor the church, according to Reinhold. Rheticus relied on Daniel's foretelling that the Son of God would defend their assemblies. All the more reason, according to the dean, for the young not to cast away these studies in despair of rewards or salvation but to study even more enthusiastically so that their knowledge may be preserved even "in the midst of enemies, if that is what destiny brings."[158] In another speech, he referred to what he called their duty to protect the embellishment of the church. He concluded grandiloquently, exhorting his readers as follows: "let us acknowledge that God will not fail His army, nor will He permit that the heavenly teaching be utterly destroyed."[159] It was this divine promise that likewise gave Reinhold confidence that his scholarly efforts were not in vain. To the contrary, the astronomer saw himself as instrumental in the transmission and preservation of the arts, even in the face of the dangers posed by the Turkish enemy. The divine promise prompted him to wage his scholarly war even more passionately. Just as Luther was prompted by the divine promise to persist in faith despite of the terrible state of the world, the astronomers persisted in their learned activity and in waging their scholarly war:[160]

---

157 *Theoricae novae planetarum*, A5^{v}: "Sed cum Deus promiserit sibi Ecclesiam suam curae futuram esse, non poterunt eam Turci funditus delere, reliquum erit aliquod domicilium doctrinae coelestis et aliarum honestarum artium, erunt aliqui mediocres principes, qui horum studiorum reliquias utcunque foveant. Ideo nondum frangamur animis, nondum de Ecclesia, de studiis prorsus desperemus."

158 Rheticus, "Oratio de astronomia et geographia," A3^{v}: "[…] inter medios hostes […] si ita ferent fata […]."

159 Rheticus, "On Natural Philosophy," 138; idem, "Oratio de physica," [B3^{r}]: "[…] statuamus Deum suo agmini non defuturum, nec permissurum ut funditus deleatur caelestis doctrina."

160 Oberman, *Luther*, 76–77, 253.

> Even though a great revolution of all empires hangs over human affairs, it does not suit magnanimous spirits to put aside the care and study of truth and virtue. No, on the contrary, because we fear that the barbarians are imminent, we fight more passionately so that the best and most useful arts survive for posterity [...]. Thus we too strive to hand down our teaching and illustrious books to others, by whose diligence and faith they are preserved for posterity.[161]

The divine promise gave meaning to their unwavering efforts to write on their subject and teach astronomy for the sake of the generations that would follow. The astronomers, moreover, encouraged the young to take up their own courses of study, and were quick to remind kings and fellow scholars of their responsibility for the preservation of learning and piety. Their warlike spirit was nurtured by God's promise that the church would survive.

---

161 *Theoricae novae planetarum*, [A8^{v}]: "Etsi enim magna impendet rebus, humanis omnium Imperiorum conuersio, tamen non propterea veritatis et virtutis curam ac studium deponere generosas mentes convenit. Imo quia barbariem impendere metuimus, pugnemus acrius, ut optimae utilissimaeque artes ad posteros perveniant [...]. Ita nos quoque contendamus aliquibus libros illustratos et doctrinam tradere, quorum industria et fide custodiantur ad posteritatem."

CHAPTER 2

Luther's Anthropology

On 17 November 1545, Martin Luther completed what would prove to be the longest work of his academic career and would later prove to be the first in a tradition of major Protestant Biblical commentaries.[1] Based on ten years of his lectures as professor *Lectura in biblia* at Wittenberg University, this was his written commentary on the first book of the Old Testament. His comments on the first eleven chapters of Genesis had already been printed in one volume in 1543, this being the only edition of which Luther himself supervised the composition and revision before his death in 1546.[2]

Moses, according to Luther, related his account without allegory.[3] By means of such a theological recovery of the history of the Old Testament, Luther distinguished himself—and Reformation theology in his wake—from the mainstream medieval hermeneutical tradition. His emphasis on the historical value of Holy Writ was in keeping with late-medieval monastic schools like the Augustinians who favored a literal-historical approach to Scripture.[4] Contemporary humankind was believed to be the offspring of the first parents and also to be living in the historical aftermath of original sin. Adam's Fall was believed to have profoundly impacted not only the relationship between God and man, but also to have affected the human condition in a radical and irrevocable manner. In order to understand the theological anthropological articulations in cosmological debate, it is crucial to fully understand, in epistemological terms, Luther's appraisal of the gravity and severity of original sin. Inevitably, his fierce opposition to the rationalist theology of the medieval Schoolmen comes to mind. His well-known portrayal of reason as the devil's whore does not altogether cover the full breadth of his thinking on human reason. In addition to this very negative image, a positive role for reason was also to be played in relation to faith.

1 Luther, "Enarratio," vol. 42.; Lull, "Luther's Writings," 48; Maxfield, *Luther's Lectures*, 23. This was not Luther's first commentary on Genesis. An earlier one, drawing on related sermons by him, was published in 1523–1524; on this earlier commentary, see: Arnold, "De la Création," 89. Philippe Büttgen remarks that the authenticity of Luther's lectures on Genesis is, among other things, validated by his preparatory notes on the first three verses of the first chapter, which are preserved at the library of the University of Jena; see: Büttgen, *Luther et la philosophie*, 91.

2 Bell, "Man," 160.

3 Luther, "Enarratio," 42: 4–5.

4 Benin, *Footprints of God*, 186.

 | DOI:10.1163/9789004387751_004

Once we have discussed Luther's appraisal of the impact of the Fall on the human condition in general, his specific take on the human capacity for knowledge about God's Word and Works will be discussed and compared to Scholastic thought. Thereafter, we consider Luther's recommendations as to the correct attitude to be adopted by the philosopher when inquiring into divine creation, hereby identifying the particulars of piety in a Lutheran learned setting.

1 The Apple of No Return

In his commentary on Genesis, Martin Luther painted a dark portrait of the postlapsarian condition of man. Condemned to ineluctable sinfulness, man was in constant need of God's Word to be reminded of His existence, and his own status as being divinely created. By successfully tempting Adam and Eve to eat from the Tree of Knowledge, Satan had struck a first blow in a resumption of his battle against God. Fought by means of unleashing waves of sin and unrighteousness, this was a fight over the true church.[5] Man's sinful state was witnessed by his perpetual attempts to assume for himself a position on a level comparable to that of God.[6]

According to Moses, God created man in His image and likeness on the sixth day of creation.[7] In an extensive commentary on Genesis 1:26 ("And God said, Let us make man in our image, after our likeness: and let them have dominion over the fish of the sea, and over the fowl of the air, and over the cattle, and over all the earth, and over every creeping thing that creepeth upon the earth"), Luther explained that Adam and Eve, before their expulsion from the Garden of Eden, had lived in a state of wonderful safety, without fear of death or danger. Being created in the image of God implied both spiritual and physical perfection.[8] The Reformer contrasted his view of the *imago Dei* with that set forth by Augustine of Hippo (354–430 CE) and other Church Fathers. Augustine defined the image of God in terms of the rational powers of the soul: namely, memory, intellect, and will.[9] Luther went further, extending the divine image from a spiritual effect to Adam's physical being, his life, and his relation both to

---

5 Maxfield, *Luther's Lectures*, 150; Oberman, *Luther*, 23, 77–78.
6 Wriedt, "Luther's Theology," 92.
7 Luther, "Enarratio," 42: 41–51 (Commentary on Gen.1:26).
8 Bell, "Man," 163.
9 Ibid.

God and to divine creation. All senses were pristine as well: "I think that before sin, Adam's eyes were sharper and clearer than those of the lynx and the eagle [...]."[10]

Despite their perfect condition, the first parents were nevertheless on the brink of original sin, of committing the sin of hubris. Luther compared their pristine condition of first innocence with early childhood when sin lies dormant inside man.[11] God had created the first parents in an intermediate state of being which combined qualities both animal and angelic. Like animals they were subject to bodily functions such as eating and drinking, while the faculty of intellect was one they had in common with the angels. Adam and Eve thus found themselves at the parting of two ways. Either they would evolve into a state of perfect innocence (a condition of immortality in which they could no longer sin and all bodily functions would have disappeared)[12] or they would lead a physical life (a condition of mortality ruled by malediction, sin, and death).[13] The animal side of this dual nature would incline man to sin.[14] Despite their "brilliantly illuminated reason," their "will conforming to the Word and the will of God," and their "nobility of body,"[15] Adam and Eve disobeyed their creator and ate the fruit of the Tree of Knowledge at the instigation of the snake.[16] With that, their fate was sealed. The first parents and their offspring were thereafter condemned to lead a sinful and physical life, ruled by ignorance, toil, and pain. Agricultural labor, once a perfect pleasure carried out in obedience to God, was now become a strenuous travail with little respite and often without yield.[17] Adam after the Fall had become no more than the corpse of the first man.[18] Luther compared the effects of the Fall to leprosy:

> But, after the Fall, like leprosy, death invaded all senses, to the extent that our intellect itself is incapable of conceiving that image [of God] [...]. On each side, one perceives neither intact will nor reason, but rather a madness more than bestial [...]. Today, how much greater are the boar's hearing, the eagle's eyesight, the lion's strength than the powers of man?

10 Luther, "Enarratio," 42: 46–47 (Commentary on Gen.1:26): "Plane enim existimo ante peccatum Adae oculos ita fuisse acutos et claros, ut lincem et aquilam superaret [...]."
11 Ibid., 42: 351 (Commentary on Gen.8:21).
12 Ibid., 42: 42 (Commentary on Gen.1:26).
13 Ibid., 42: 84 (Commentary on Gen.2:16–17).
14 Ibid., 42: 85 (Commentary on Gen.2:17).
15 Ibid., 42: 106 (Commentary on Gen.3:1).
16 Ibid., 42: 73 (Commentary on Gen.2:9).
17 Ibid., 42: 29 (Commentary on Gen.1:11).
18 Ibid., 42: 106 (Commentary on Gen.3:1).

> No-one can imagine how much greater his nature was then than it is now [...]. But if you happen to sin, you lose this image and you die. That is why, today, we see what great dangers, how many deaths and mortal threats our miserable nature is forced to endure. We will never be safe in God: fear and terror torment us even in our sleep. These evils, and similar ones, are the image of the devil, who marked us with them.[19]

Thus marked by the image of the devil, sin pervading the entirety of their being, humans were no longer capable of understanding the divine image, nor of what the divine image consisted. The faculties of will, memory, and intellect persisted, but were irreversibly defiled and weakened: their functioning had become permanently impure and leprous.[20]

Adam's Fall had changed man's relation to God and His creation dramatically and irreversibly. While original sin predisposed man to commit corporeal sins due to "the madness of our desires," the sins of the soul were the most egregious.[21] The soul became capable of committing sins previously unknown to Adam. Transgressions like disbelief, ignorance of God, desperation, hate, and blasphemy now became part of the lot of mankind. As a comment on a genealogical table from Adam to Abraham, Luther wrote that this particular line of descent mirrors the human condition: creatures so deformed by sin that they could neither know their own origin nor God himself, were it not for the Word of God revealing these "glimmers" of divine light to us from afar.[22] Scripture, however, could prompt the beginning of the renewal of the divine image in a threefold manner: first, by having faith, man is reborn into the hope of leading an eternal life with God; secondly, faith can inspire man to exert himself to obey God, as instructed by the Word and aided by God; finally, God's Word can make man more resistant to the committing of sins like incredulity. Full

---

19 Ibid., 42: 46–47 (Commentary on Gen.1:26): "Sed post lapsum irrepsit mors tanquam lepra in omnes sensus, ita ut ne intellectu quidem imaginem istam possimus assequi [...] Utrinque igitur nec ratio nec voluntas integra cernitur, sed furor plus quam beluinus. [...] Apri quantum hodie audit, aquilae quantum visu, leo quantum viribus hominem antecellit? Quare nemo cogitare potest, quanto melior natura tum fuerit, quam nunc est [...]. Si autem peccaveritis, amittetis hanc imaginem et moriemini. Nunc igitur videmus, quanta pericula, quot mortes et mortis occasiones misera haec natura cogatur experiri et sustinere praeter foedam concupiscentiam et alios furores peccati ac inordinatos motus, qui in omnium animis fiunt. Nunquam sumus securi in Deo, terror et pavor etiam in somnis nos exercent. Haec et similia mala sunt imago Diaboli, qui ea nobis impressit."

20 Ibid., 42: 86 (Commentary on Gen.2:17).

21 Ibid.

22 Ibid., 409 (Commentary on Gen.10:32).

restoration of the divine image, however, is only possible in life after death.[23] After man's earthly death, Luther believed that God would resuscitate him to an eternal and new life.[24] The German Reformer and his followers were anxiously looking forward to the second coming of Christ.[25] On this Last Day, all deceased bodies would be resurrected and reunited with their souls. Christ would then judge every individual, and the present world would be fully transformed and destroyed. A new heaven and new earth would be the setting for the "incomprehensibly perfect Kingdom of God, in which the faithful would dwell eternally with Christ."[26] Man's posthumous condition was to be even more spiritual than Adam's pristine condition of innocence: man will attain a state of perfect innocence, characterized by immortality and the absence of sin.[27]

Luther's anthropology is grafted onto the doctrine of the immortality or eternity of the soul. Original sin had not made the human soul mortal; only his physical body. The immortal soul proved that God had created humankind for a specific end, that being a spiritual and eternal life. Genesis was to reinforce the human hope of immortality that resonates with the presence of the divine image in man and which differentiates him from the other creatures.[28] God, on the Last Day, would raise all men from the state of death to a new form of existence, eternal life.[29]

In spite of the incipient renewal of the divine image inside man, the impetus for salvation, the process towards achieving the state of blessedness, could not be realized by man: only an act of divine grace could remove the sinful condition of man. In general, grace is to be understood as God's assistance, support, and favor of humans.[30] Due to the loss of their pristine integrity, Luther believed that man's natural qualities could not save himself in any way.[31] The Lutheran view of man's inability to effect his own salvation departs considerably from Catholic theology, which taught that man could influence his salvation by the performance of good works or other acts of penance and virtue. In Reformed opinion, the Fall had resulted in a situation where only faith in Christ—*Sola Fide*—and only divine grace—*Sola Gratia*—could open the way to salvation.[32]

---

23 Ibid., 48.

24 Ibid., 148–149.

25 Oberman and Barnes situate Luther's eschatological preoccupations in the late-medieval millenarist tradition; see: Oberman, *Luther*, 68; Barnes, *Prophecy*, 30.

26 Barnes, *Prophecy*, 36–38.

27 Luther, "Enarratio," 42: 76.

28 Ibid., 42: 66 (Commentary on Gen.2:7).

29 Tiililae, "Resurrection," 3: 2048.

30 Hillerbrand, "Grace," 94.

31 Luther, "Enarratio," 42: 45 (Commentary on Gen.2:71).

32 Crowther, *Adam and Eve*, 2–3; Kusukawa, *Transformation of Natural Philosophy*, 29.

Luther's anthropology radicalized the Catholic perception of the sinful condition of man. The sin of hubris became deeply rooted in theological anthropology. Sinfulness became an ineluctable state of being for postlapsarian man: "human nature by itself was incapable of casting off sin, of escaping the castigation of sin and death and of regaining the lost obedience."[33] Pride caused man to disdain God and to no longer fear his creator. Luther proclaimed that one needed to take in the full extent of the consequences of the Fall, and, adopt a humble attitude before God.[34]

> This manifold corruption of our nature, therefore, should not be minimized, but rather treated at length, for man has fallen from being an image of God, from the knowledge of God and of all other creatures, from his honorable nakedness, into blasphemy, into hate, into the contempt of and, still worse, hostility towards God! I haven't said anything yet about the tyranny of Satan, to whom our miserable nature is subordinate due to sin. One must dwell on these things, I repeat, because if they are not rightly recognized, the remedy will be neither known nor desired. The more sin is minimized, the more grace itself is degraded.[35]

Humility was the key to the acknowledgment of one's sinful condition. In Luther's opinion, the cause and consequence of original sin was that neither Adam nor his offspring had been capable of employing the divine gifts in a humble manner. Unable to restrain themselves, man acted haughtily before God.[36] A gift is to be understood as something superadded by God to the natural being of man: it is the gift of grace in addition to man's natural condition.[37] Out of goodness, Luther wrote, God granted humankind the divine gifts of an intact

33 Luther, "Enarratio," 42: 146–147 (Commentary on Gen.3:15): "[…] satis clarum est naturam humanam post lapsum suis viribus nec peccatum tollere nec poenas peccati et mortem effugere nec obedientiam amissam recuperare potuisse."

34 McGrath, *Luther's Theology*, 154.

35 Luther, "Enarratio," 42: 107 (Commentary on Gen.2:17): "Non itaque haec naturae multiplex corruptio extenuanda sed magis amplificanda est: Quod homo ab imagine Dei, a noticia Dei, a noticia aliarum creaturarum omnium, ab honestissima nuditate in blasphemias, in odium, in contemptum Dei, Imo quod plus est, in inimicitiam erga Deum lapsus est. Taceo iam de tyrannide Satanae, cui propter peccatum misera haec natura subiecta est. Amplificanda, inquam, haec sunt, propterea quod, nisi recte cognoscat magnitudo morbi, remedium quoque non cognoscitur nec desideratur. Quanto enim magis peccatum extenuaveris, tanto quoque gratia magis vilescet."

36 Ibid., 42: 264, 265.

37 Bell, "Man," 164.

body, wisdom, royal power, knowledge of the arts, and physical strength.[38] If pride was shown before these gifts, God would humiliate and humble man by making him experience His wrath, by the threat of hell and eternal damnation, by temptation and suffering.[39] One of the most dramatic manifestations of the divine wrath after the Fall was the universal Flood that destroyed the original world and all human beings, sparing only Noah and his family.

2 Knowledge after the Fall

What are the epistemological consequences of the Fall? Markus Wriedt explains how the sinful condition of man, in Luther's theology, "becomes most manifest in relationships with authorities from outside the Bible such as Aristotle and a Scripture interpretation led astray by church traditions lacking true theological back-up."[40] Hence Luther's characterization of reason as the devil's whore and his objections and insulting remarks about the rationalist theology of the Scholastics. While such statements have led to his being characterized as a fideist who deeply mistrusted human reason, a more balanced picture will be presented here. Denis Janz and Robin Barnes have contended that Luther also reserved a positive role for reason in relation to faith; both emphasize that the Reformer, despite his clear mistrust of human reason, considered it a divine gift to mankind.[41]

"Original sin," wrote Luther, "is indeed a total destruction of human nature: the intellect darkened to such an extent that we no longer recognize God and His will, that we do not discern the works of God."[42] He thought of Adam and Eve as excellent philosophers in their prelapsarian condition, for they were entirely free from sin.[43] A perfect knowledge of God implied a perfect knowledge of all creatures and everything inferior to God.[44] By contrast with Thomas Aquinas (1225–1275), who is taken here as the spokesman of Scholastic theology on the subject of the abilities of reason as set forth in his *Summa contra gentiles* (c. 1258/59–1264), Luther did believe that the Fall had deeply affected

38 Luther, "Enarratio," 42: 265.

39 McGrath, *Luther's Theology*, 154.

40 Wriedt, "Luther's Theology," 92.

41 Janz, "Whore or Handmaid," 47; Barnes, *Prophecy*, 8.

42 Luther, "Enarratio," 42: 86 (Commentary on Gen.2:17): "Sed peccatum originale est vere totus lapsus naturae humanae, quod est intellectus obscuratus, ut non agnoscamus amplius Deum et voluntatem eius, ut non animadvertamus opera Dei."

43 Ibid., 42: 49 (Commentary on Gen.1:26).

44 Ibid., 42: 106 (Commentary on Gen.3:1).

the human intellect.[45] According to Thomas, who distinguished natural from supernatural gifts, Adam's Fall caused the loss of man's supernatural gifts while his natural gifts such as reason remained untouched and hence, the powers of reason were still adequately functioning after the Fall.[46] Thomas did not think that reason required supernatural intervention to operate properly; it was a natural faculty. Luther, by contrast, deemed human ratio no longer capable of functioning with the same integrity after the Fall. In railing against the Scholastic theologians, Luther contrasted their respective positions as hubristic versus expressive of humility.

> Those [i.e., the Scholastics] who minimize this evil [due to original sin] and who speak about corrupted nature as if it were not corrupted, in the manner of philosophers, are the cause of great errors. Thus they say that the natural gifts have remained intact [...]. We see properly and we are conscious of the little that is left to us. But the disputers who affirm the integrity of natural gifts do not take into account all things we have lost. Because, gone are the will, good and just, which is pleasing and obedient to God, which confides in its Creator and which, while giving thanks, makes good use of creation. To this extent, we want to see the devil where God is and that the mere mention of His name submerges us in fear, especially because the thought about His judgment overwhelms us. Is this what is meant when talking about our intact nature?[47]

Those that speak about corrupted nature in the manner of the philosophers are identified a few paragraphs later as the Scholastic theologians. The Schoolmen were oblivious to the deteriorating consequences of the Fall. Reason, Luther wrote, is overcome by sin, is always under the wrath of God, and cannot extricate itself from this evil by its own means. Scripture shows that what reason contrives is malicious, impious, and sacrilegious.[48] Scholastic theology, in

45 te Velde, "Natural reason," 117–140.

46 Harrison, *Fall of Man*, 43–44, 56.

47 Luther, "Enarratio," 42:106 (Commentary on Gen.3:1): "Magnorum itaque errorum causam praebent, qui hoc malum extenuant et loquuntur de natura corrupta Philosophorum more, quasi non sit corrupta [...] Quid et quantillum remanserit, aliquo modo videmus et sentimus. Sed quantum amiserimus, profecto non vident, qui disputant naturalia mansisse integra. Nam voluntas bona, recta, Deo complacens, Deo obediens, fidens in Creatore et creaturis recte cum gratiarum actione utens amissa est, Adeo ut ex Deo nostra voluntas faciat Diabolum et horreat ad mentionem nominis eius, praesertim cum premitur iudicio Dei. Hoc num, quaeso, est naturalia esse integra?"

48 Ibid., 42: 348 (Commentary on Gen.8:21).

Luther's opinion, had degenerated into a philosophy that offers no true knowledge of God and "moves in shadows" precisely because it had turned away from God's Word.[49] Luther considered the Scholastic theologians as prisoners to their personal philosophical imagination who had failed to hand down true knowledge about God or themselves.[50] Instead, the Reformer suggested a humble reason in obedience of God's Word. Luther argued that education could serve as a refinement of reason's impulses to sin, and have a moderating and tempering effect on the young mind. One must not overestimate the wonders of education though, he argued, since to speak about virtue and vice, even when it comes to great men who temper and control their impulses to sin, is to ignore man's malicious nature. The different manifestations of evil can never be fundamentally removed from human nature.[51]

In line with Scholastic theology, Luther ruled out the possibility of an immediate knowledge of God while man resided on earth, a state that would come to an end after the general Resurrection.[52] After death, individual human beings would encounter God face-to-face. During the period before this beatific vision, Christians relied on intermediaries to acquire a mediated knowledge of their creator.[53] Luther conceived of God as incomprehensible to man; He was hidden to humanity within his self-revelation (the principle of *Deus absconditus*).[54] God revealed himself through His works and His Word, things that were within the sensible and intellectual reach of man. While touching upon the unknowability of God and his whereabouts before creation, the Wittenberg theologian argued that even after the moment of creation, God would still be fully incomprehensible if he had not revealed himself through the visible manifestations of his Word and creation (the principle of *Deus revelatus*).

> Even now, after creation, He is present within, without and above all creatures, that is, He is still incomprehensible [...]. This is also the reason why God manifests Himself only in His works and in His word, because these

49 Ibid., 42: 349–350 (Commentary on Gen.8:21).

50 Kusukawa, *Transformation of Natural Philosophy*, 44, 201, 347–349.

51 Luther, "Enarratio," 42: 352 (Commentary on Gen.8:21). Luther relates the intensity of the human inclination to sin to the different stages in life. Similar to Adam's state of first innocence, the propensity to sin is dormant in children. Sin shows itself from youth onwards and grows stronger with age. In this respect, he reminds his readers of the barbaric verse "angelical youth, satanic old age."

52 Barnes, *Prophecy*, 7.

53 Kusukawa, "Introduction," xxiii; te Velde, "Natural reason," 135.

54 McGrath, *Luther's Theology*, 149, 191.

> things are comprehensible in some way, and the rest, which are characteristic of divinity, cannot be grasped or understood as they are [...].[55]

In this commentary, God's works or creation emerge as another locus of divine revelation. The postlapsarian relation between humanity and God is constituted by means of the former's rational engagement with the latter's visible manifestations in this world. The present distinction between Holy Writ and the Book of Nature coincides with the concept of a twofold knowledge of God: namely, God as Redeemer and God as Creator. This is in fact a Scholastic division, crediting human beings with two ways to achieve cognition about God: *per revelationem* and *per creaturas*.[56] Certain aspects of divine being would remain permanently hidden from men. As such, God remained fundamentally hidden, and, therefore, effectively unknown.[57] Out of the two divine visible manifestations in the world—the Book of Nature and the Holy Scripture—it was the latter that occupied a position of primacy in constituting a temporal relation between God and humankind. In spite of their postlapsarian depravity, human beings retained some inborn notions about the existence of divinity but these notions were "nearly extinct."[58] Scripture was needed before man was able to consider nature as a second book authored by God.[59] Safely channeled through revelation, these natural *notitiae* of God produced certain insights regarding the existence of divinity,[60] but none regarding God's will as would have been made known by revelation. The light of natural reason functions to a limited extent but halts at the cognitive level; only faith may induce existential insights regarding divine will.[61]

2.1 *God's Word*

Luther thought of Scripture as having a soteriological purpose, instructive on the subject of divine will: "The end that Scripture reveals to us, is, above all, that man has been created in the likeness of God to live with Him in eternity, and to preach God here on earth, to give thanks to him, and to obey his Word

---

55 Luther, "Enarratio," 42: 9 (Commentary on Gen.1:2): "Nunc autem post creationem esse intra, extra et super omnes creaturas, hoc est, etiam esse incomprehensibilem [...]. Ideo Deus quoque se non manifestat nisi in operibus et verbo, quia haec aliquo modo capiuntur; reliqua, quae propria divinitatis sunt, capi aut intellegi non possunt, ut sunt [...]."

56 te Velde, "Natural Reason," 134.

57 McGrath, *Luther's Theology*, 165.

58 Luther, "Enarratio," 42: 50 (Commentary on Gen.1:26).

59 Arnold, "De la Création," 21.

60 Janz, "Whore or Handmaid," 48–49.

61 McGrath, *Luther's Theology*, 162.

in submission [...]."[62] The readability of Scripture was essential. Luther's view on the appropriate way to read and understand Holy Writ was one of the main fault lines with Scholastic theology. The proper interpretation of Scripture was crucial against the background of early modern confessionalization.[63] Luther sought to indemnify the reading of Scripture from the illicit presence of human dogma, as represented in the use of Catholic tradition and Aristotelian philosophy in deciphering the meaning of Biblical passages. Holy Scripture was the sole source of Christian faith (*Sola Scriptura*). Luther's Bible-oriented theology would imply a significant change in the task of the theology professor, who was to be principally concerned with Biblical exegesis.[64]

Medieval Catholic theology had never denied the supreme authority of Scripture but the church was perceived as the final and infallible judge in determining the true meaning of passages.[65] The hermeneutical authority of the church was reaffirmed by the fourth session of the Council of Trent on 8 April 1546.[66] Luther's hermeneutics were grounded on the principle of *claritas Scripturae*. Thanks to the perspicuity of Scripture in matters of faith, the individual Christian could have immediate access to theological doctrines in God's Word on the condition that he had been illuminated by God (inner clarity). The conscience would be compelled to believe what is true.[67] Luther's advocacy of the perspicuity of Scripture was the Reformed answer to the Catholic belief in the obscurity of Scripture. He equated this clarity with the literal or historical sense of the Bible.[68] Traditionally, medieval hermeneutics consisted of four distinctive senses, the quadriga. The literal sense served as a mere stepping stone to the spiritual level of interpretation, consisting of the allegorical, moral, and anagogical senses. Luther distanced himself for the major part from medieval allegorizing and pushed the literal sense to an unprecedented degree.[69] The use of what, to him, seemed preposterous allegories had often garbled the true historical meaning.[70] Biblical passages were either clear by themselves, or their

62 Luther, "Enarratio," 42: 98 (Commentary on Gen.2:21): "Principalis igitur finis est, quem scriptura ostendit, quod homo sit conditus ad similitudinem Dei, ut cum Deo in aeternum vivat et hic in terra Deum praedicet, ei agat gratias, et verbo eius obediat in pacientia [...]."

63 Methuen, "Interpreting the Books of Nature and Scripture," 180.

64 Brockliss, "Curricula," 2: 595–596.

65 Kraeling, *Old Testament*, 9.

66 Howell, *God's Two Books*, 23–24; Kraeling, *Old Testament*, 33.

67 Callahan, "Claritas Scripturae," 355; Kraeling, *Old Testament*, 9–10.

68 Callahan, "Claritas Scripturae," 363.

69 Preus, *From Shadow*, 143.

70 Luther, "Enarratio," 42: 377.

obscurity could be resolved by comparison with clear passages. The theologian emancipated the literal sense, so to speak.[71] Throughout the sixteenth century, in general, there was a clear move towards the literal interpretation of sacred texts, both in Protestant as in Catholic circles.[72] Although there was a renewed interest in the literal sense in Catholic circles, the *sensus litteralis* did not prevail in Biblical exegesis as it had in Protestant exegesis.[73] In Luther's view, no external factors besides Scripture were needed to understand Biblical text in its literal sense (the principle of outer clarity).[74] He wished to exclude the Scholastic use of rationality from the domain of faith.[75] Unlike Thomas, he saw no place for philosophy or rationalist argumentation in the reading of God's Word, given what he regarded as the insufficiency of the human intellect. Biblical verses do not require rational comprehension so much as acquiescence.

Beyond the Bible itself, Luther likewise objected to philosophical interference with theology. Aristotle was not a suitable authority in matters of faith. Tenet 50 in his 1517 *Disputatio contra scholasticam theologiam* stated that Aristotle was to theology what darkness was to light.[76] To Luther, Scholastic philosophy brought about a deformation of God's Word.[77] The manner in which Scholastic theologians employed Aristotelian philosophy to explain Scripture led, to quote Philippe Büttgen, to "a philosophical contamination of the Word," the *schola* or university being the carrier of the virus.[78] In his comment on Genesis 1:14 regarding the creation of the heavenly luminaries, Luther gave some examples of illicit rational behavior by theologians. He especially rejected the tendency to explain philosophical knowledge of the heavens theologically. There are some theologians, Luther wrote, who designate the supracelestial waters that Moses described in Genesis 1:6 as the crystalline sphere. Luther thought of this as "childish." He himself admitted that he did not understand Moses's words rather than approving these products of ignorance.[79] The intellect was no longer sufficiently equipped to rationally comprehend Scripture.[80]

---

71 Harrison, *Bible*, 107.

72 Ibid., 70.

73 Howell, "Hermeneutics of Nature and Scripture," 279.

74 Callahan, "Claritas Scripturae," 356.

75 Kusukawa, *Transformation of Natural Philosophy*, 33.

76 Büttgen, *Luther et la philosophie*, 60.

77 Ibid., 10–11.

78 Ibid., 16.

79 Ibid., 22.

80 Some differentiation regarding Luther's hermeneutics is in order. Not the whole of Scripture was thought of as perspicuous; neither was every single human being believed to

2.2 *God's Works: The Book of Nature and the Heavenly Lights*

Despite his vehement attacks on the Scholastic use of reason, Luther did think that reason could be appropriated positively in relation to faith in terms of a supportive role. The Wittenberg Reformer explicitly depicted the contemplation of nature as witnessing divine providence. His commentary on Genesis indeed shows that rational engagement with divine creation, when guided by Scripture, could lead to the recognition of God as the creator of the visible realm. Rather than a rationalist argumentation, Luther seemed to consider the visible realm as a place where God and His providence could be recognized and admired rather than rationally proven. The whole of the world we can see was a testimony to divine providence for it was created by God as told by Moses in Genesis.[81] Contemplating the Book of Nature implied contemplating the visible effects of divine providence and admiring God as the final and efficient cause of his creation in his creation.[82] Illuminated by faith, "aided" reason applied to the Book of Nature was, as Robin Barnes has contended, a source of Lutheran piety.[83] "Is it not our task," Luther asks the reader of his commentary on Genesis, "to admire His works and to sing their praise always and everywhere?"[84] Observation and contemplation of divine creation is the worthiest preoccupation of man, wrote Luther.[85] Admiration progressively strengthened faith.

Luther mentioned another example of witnessing providence when writing about the different species on earth. To him, the fact that identical members of a species are born out of other members of that species was a token that the world did not come into existence coincidentally. This order reflected divine providence.[86] Luther's examples regarding the admiration of divine creation

---

be able to read it properly, nor was Scripture exclusively consulted in Biblical exegesis. Holy Writ was believed to remain obscure in certain matters due to the inevitably sinful condition of the reader and the permanently hidden character of God. The weakness of the reader was an integral part of his own understanding of the Bible. Furthermore, the principle of *Sola Scriptura* did not imply that glosses or commentaries were no longer appropriate. Finally, as exemplified in the commentary on Genesis, Luther added glosses and explanatory notes to his translations of the New and Old Testament to guide the correct interpretation; see: Crowther, *Adam and Eve*, 34; Maxfield, *Luther's Lectures*, 43.

81 Luther, "Enarratio," 42: 28 (Commentary on Gen.1:11).

82 Ibid., 42: 95 (Commentary on Gen.2:21).

83 Barnes, *Prophecy*, 8–9.

84 Luther, "Enarratio," 42: 95 (Commentary on Gen.2:21): "Si enim credimus, causam efficientem et finalem esse Deum, an non opera eius admirari, iis delectari et ea semper et ubique debebamus praedicare?"

85 Ibid., 42: 24 (Commentary on Gen.1:6).

86 Ibid., 42: 28 (Commentary on Gen.1:11).

are but a few of many that attest the guise of religious observance to the observation of nature in a Lutheran context.[87]

Luther's comment on Genesis 1:14, covering the creation of the heavenly lights on the fourth day, gives a concrete illustration how the observance of nature could invoke piety on behalf of the observer (in the broader sense) of heaven. In his comments, Luther is seen to join ranks with an entire sixteenth-century tradition taking up Genesis 1:14 as illuminating the legitimacy of astronomy and astrology. Historiography has mainly focused on Melanchthon's use of the quote whereas Luther's comments have been neglected. Luther's comment on the first sentence of the verse explains the difference between the astronomical and his own interest regarding the contemplation of the stars. Whereas the astronomers discuss the size of the celestial bodies and take the view that both the stars and moon receive their light from the sun, Luther claims to neither affirm nor condemn such views. The theologian, however, looks to the celestial bodies as divine creations that came into existence for the use of man.

> It is sufficient to me that we recognize the benevolence and power of God in the beauty of the celestial bodies and in their utility to our lives, because it is through His Word that He has created such great things and that He preserves them for our use still today. These things pertain to our profession, that is, theology, and have the power of strengthening souls.[88]

Observance of the effects of divine creation was advanced as a source of piety and as being capable of strengthening faith. Creation led the faithful contemplator to the creator. Luther's comments in the second sentence clarify the specific utility of the celestial bodies for humankind on earth, bringing his readers to the field of astronomical practice. Based on his reading of the verse in question, he distinguishes two purposes. First, God intended the heavenly lights to announce His wrath or calamities yet to unfold. These signs concerned extraordinary occurrences such as eclipses and major conjunctions of the stars.

87 The term observance is borrowed from Katherine Park who argues for the spiritual dimension of twelfth-century monastic observations of the stars. The stars were primarily objects of religious contemplation: through contemplation of divine creation, the minds of the pious observers were drawn toward God; see: Park, "Observation in the Margins," 23.

88 Luther, "Enarratio," 42: 31 (Commentary on Gen.1:14): "Mihi satis est, ut in istis corporibus tam elegantibus et nostrae vitae utilibus cognoscamus et benignitatem Dei et potentiam, quod tantas res verbo condidit, et adhuc hodie conservat ad nostrum usum. Haec nostrae professionis, hoc est, theologica sunt, et valent ad animos confirmandos."

"If this might sound too primitive," Luther added, "let us not forget that Moses wrote for an uncultured people."[89] Meteors, rainbows, and other aerial phenomena equally counted as signs; the distinction between celestial spheres was not made in Scripture that speaks only of heaven and earth. Heaven emerged here as a theater that displays signs of future events to humanity, hence as a medium that made divine providence visible to earth-bound humanity. The second purpose of the heavenly lights, according to Luther's interpretation of Holy Writ was to mark the passing of time, meaning seasons, years, and days. Counting the days and years, Luther wrote, was "a good deed that reveals creation and the order intended by God."[90] The observance of creation leads to the creator Himself and His providential rule of the universe in terms of time.

From a theological anthropological perspective, Luther credited the acts of counting and measuring, thus the mathematical disciplines, as activities that made man conscious of his religious being. He wrote that the mathematical disciplines were a divine gift to humankind that revealed the immortality of the human soul which distinguished man from other created beings on earth. Man alone can count time.[91] Luther linked the significance of man's natural tendency to count with the end of human existence: to comprehend the motion of heaven or to measure the celestial bodies demonstrates that man is a creature made to live in heaven and not, like the animals, in the infirm part of the universe. In this act, "a spark of eternity shoots forth," making man realize that in the end he belongs to heaven.[92] The pursuit of astronomy enabled the virtuous astronomer to affirm his immortal nature (i.e., the part destined for eternity and divinity) and the part of him that was still godlike after the Fall: "so that with the help of the mathematical disciplines, which no one can deny have been divinely revealed, man may fly up in spirit and, leaving behind those things that are earthly, consider and examine heavenly things."[93] These transcendent capacities revive a Platonic metaphor of the ascent of the soul. By means of the mathematical disciplines that God had revealed to man, the

89 Ibid., 42: 32 (Commentary on Gen.1:14): "Quod si hoc crassius videtur, meminerimus Mosen rudi populo scripsisse."

90 Ibid.: "Quod igitur dies numeramus, item annos, beneficia sunt creationis et ordinationis divinae."

91 Ibid., 42: 33 (Commentary on Gen.1:14).

92 Ibid., 42: 34 (Commentary on Gen.1:14).

93 Ibid., 42: 34–35 (Commentary on Gen.1:14): "[…] ut mathematicarum disciplinarum adminiculo, quas divinitus ostensas esse nemo potest negare, homo ex terra in altum animo subvolet, et omissis, quae in terra sunt, coelestia curet et investiget."

pilgrim or exile flies in spirit through the heavens to study the celestial bodies, leaving behind everything on earth:

> Man is thus a creature made to inhabit the heavens and, having one day left behind the earth, to live the eternal life. For this is the creature who cannot only speak and judge (which pertain to dialectic and rhetoric), but also learn thoroughly the whole domain of knowledge.[94]

Luther argued that our awareness of the eternal life to come should, more than anything else, occasion expressions of gratitude to God, for it raises man's consciousness that he is actually contemplating his homeland. Heaven is the true homeland of humankind. After this life, Luther adds, "we will understand perfectly."[95]

3 Humbling the Philosopher

The knowledge of God and of His creation were intimately linked in Luther's opinion. Before the Fall, Adam was said to dispose of a perfect cognition of the specific nature of animals, fruits, trees, and other creatures; of the stars and astronomy, and all creatures were obedient to him like Adam himself was to God.[96] As a consequence of original sin, man lost this privilege irrevocably and by no means was it to be restored in this life.[97] In line with his pessimistic appraisal of the course of history, Luther argued that Adam's knowledge was gradually obscured through its transmission among his offspring and was practically extinguished.[98] Even though some remnants might have been preserved in writing, these were but a pale imitation of Adam's original knowledge.[99]

3.1 *Metaphysical Ignorance*

Throughout his commentary on God's creation of the world and humankind, Luther developed his view on the particulars of man's rational engagement

---

94 Ibid., 42: 35 (Commentary on Gen.1:14): "Ergo est homo creatura condita, ut relicta aliquando terra coelestia habitet et aeternam vivat vitam. Hoc enim est, quod non solum loqui et iudicare possit (quae ad dialecticam et rhetoricam pertinent), sed quod etiam mathemata omnia perdiscit."

95 Ibid.: "[…] perfecte intelligemus."

96 Ibid., 42: 47–49 (Commentary on Gen.1:26).

97 Ibid., 42: 49 (Commentary on Gen.1:26).

98 Ibid., 42: 90–91 (Commentary on Gen.2:20).

99 Ibid.

with visible creation after the Fall. In terms of the attainability of knowledge of nature, Luther set forth a pessimist position. Whereas prelapsarian knowledge of nature had implied a perfect cognition of the invisible aspects of creation (the *natura* of animals and other creatures), postlapsarian reason found itself restricted to the visible realm. Impaired by original sin, the intellect stood in need of God's Word to grasp that nature was divinely created, and had to acquiesce in information about nature rendered in Genesis rather than rationally comprehending divine creation. More than anything, Scripture taught man about his pervasive ignorance and raised the awareness of his humanity: God's Word humbled the philosopher. In spite of man's intellectual deficiency and his resulting ignorance, Luther believed that the mathematical disciplines, which he perceived as divine gifts, could yet yield demonstration and sure proofs.[100]

Luther's usage of the Aristotelian notion of four causes is a convenient framework for the modern reader to clarify his views on epistemology and the knowability of nature. In general, he seemed to find postlapsarian reason bound to a condition of what I refer to as "metaphysical ignorance", implying that man is ignorant in reference to the indiscernible aspects of nature. The deficient human intellect was not capable of grasping nature beyond its visible manifestation. Luther wrote that, if not aided by God's Word, reason could not make that extra step in its valuation of creation, and was found utterly incapable of designating the final and efficient cause of nature.[101] The scope of the intellect was restricted to material and formal causes: "We cannot even show the efficient and final cause plainly, which is especially miserable, since the world in which we exist and live, and we ourselves also, are the proper subject of debate and thinking."[102]

While Adam's cognition of nature had been a divine gift, enabling him to intuitively know the right names of the animals, postlapsarian knowledge was to be gained through experience which literally took ages and was made up of diligent descriptions resulting from patient observations. Some traces of the once-perfect knowledge could therefore be found in the books of the wise.[103] This type of knowledge, by definition, was acquired *a posteriori*, from experience.

---

100 Harrison, *Fall of Man*, 100.

101 Luther, "Enarratio," 42: 93 (Commentary on Gen.2:21).

102 Ibid.: "Causam efficientem et finalem plane non possumus ostendere, praesertim quod miserabile est, cum de mundo, in quo sumus et vivimus, item de nobis ipsis est disputandum aut cogitandum."

103 Ibid., 42: 90–91 (Commentary on Gen.2:20).

> And thus it is not surprising that we have no knowledge of God, since we do not even know the natures of the animals, or what virtue and efficacy they possess. There are various books at our disposal that describe the natures of plants and animals. But how much time, and what lengthy observation was necessary, for the accumulation of these things by experience![104]

A similar quote on the "unknowability" of the nature of created things is found in Luther's comment on Genesis 1:14, claiming that because reason "is too weak to achieve such knowledge," man remains in the sphere of likelihood or probability.[105] The above quotes match the usual historiographical conclusions about Luther's appraisal of the domain of philosophy as restricted to the earthly realm.[106] Luther is argued to have maintained a distinction between two realms of human existence. Man's reason was self-sufficient in matters of "*das irdische Reich*," the "*regnum rationis*," but fails in matters of the "*regnum Christi*."[107] Luther scholars assume that he thought that philosophy applied to the earthly realm, referring matters of salvation to the heavenly realm. Given the focus on experience and observation in the above quote, it seems more appropriate to replace the opposition between the earthly and heavenly realm with the dichotomy between the visible and invisible realm, distinguishing the physical from the non-physical domain. Material and formal causes of physical phenomena fall within the domain of human sensible experience whereas the efficient and final causes are not discernible. The operation of human reason is restricted not to the earthly but to the visible realm. Replacing *das irdische Reich* with the visible realm does not alter the implications of the traditional distinction which assigns the earthly realm to human reason and the heavenly realm to salvation; the sensible scope of sight merely extends it to the heavenly phenomena.

Luther argued that theology introduced a rule "that is too little known among philosophers": even though God created everything by means of His Word, He was not a prisoner of His own creation but changed creation as He pleases. Exceptions to experience-based conclusions [i.e., miracles], are

---

104 Ibid., 42: 90 (Commentary on Gen.2:19): "Itaque non mirum est, quod Dei nullam cognitionem habemus, cum etiam bestiarum naturas, cuius virtutis et efficaciae sint, non noverimus. Extant varii libri, qui plantarum et bestiarum naturas describunt. Sed quanto tempore, quam longa observatione opus fuit, dum haec per experientiam ita colligerentur."

105 Ibid., 42: 31 (Commentary on Gen.1:14).

106 Büttgen, *Luther et la philosophie*, 93.

107 McGrath, *Luther's Theology*, 137.

therefore possible.[108] Although Luther did not explicitly refer to the Scholastic notions of divine power in the verses under discussion, his words indicate the implicit awareness of the gap between divine ordained and divine absolute power. The philosopher found himself limited to beholding the visible actualization of God's absolute power. Exhaustive knowledge of nature, meaning universal truth, would imply rational comprehension of divine omnipotence, of the invisible or indiscernible aspects of nature. Luther's explanation echoed that philosophers needed to be aware of this asymmetrical relation with God and thus of the limitations of their conclusions or conjectures, and hence should not attempt to place themselves in God's position.

3.2 *Scripture as a Lamp*

The importance of God's Word in achieving knowledge about nature derives, in Luther's opinion, from its ability to take that extra step in the evaluation of creation, going beyond the visible. Scripture taught man about the origin and end of the whole of creation and thus about the efficient and final cause of nature.[109] The whole visible realm is a testimony to divine creation. Luther however did not think of the Bible as a source of knowledge about nature, although it did render some information thereon. Philippe Büttgen characterizes his analysis of Genesis as "a non-savant commentary," marked by a simplicity of language and displays of ignorance by the Biblical exegete.[110] At the beginning of his comment on Genesis 1:14, Luther stated that Moses wrote for an ignorant and young people, and did not want to put more in writing than what was necessary and useful to know.[111] In contrast to the conjectures of human reasoning, Scripture represented true wisdom. Rather than a source of knowledge, Luther staged God's Word as a lamp to be used in avoiding the pitfalls of unaided reason. In his introduction to the first chapter of Genesis, the German Reformer immediately emphasized the limits of human reason by affirming the primacy of revelation. The Mosaic account of the creation of the world taught man that God created the world and its creatures from nothing in six days and that the world would last for six thousand years. Aristotle, by contrast, argued that the world exists eternally and will be inhabited by an infinite succession of people(s). Luther adduced that man was rationally incapable of rebutting Aristotle's statement but that if we did not understand why the world has been created in six days, "let us remain dis-

108 Luther, "Enarratio," 42: 21 (Commentary on Gen.1:6).
109 Luther, "Enarratio," 42: 98 (Commentary on Gen.2:21).
110 Büttgen, *Luther et la philosophie*, 11, 94–97.
111 Luther, "Enarratio," 42: 25 (Commentary on Gen.1:8).

ciples and leave the authority to the Holy Spirit."[112] Moses, after all, is a far better teacher. Equally on the subject of the divine creation of the world, Luther raised the specter of the Epicurean philosophers who claimed that the world and humankind had come into existence by chance: "Reason is led to these labyrinths when it is deprived of the Word and follows its own judgment."[113] The same argumentation applied to the historical fact that Eve had been created from Adam's rib while he was asleep. While to reason this was a very beautiful and amusing "fable," to us, Luther explained to his audience and readership, this wisdom revealed that the generation of man was commenced by the Word.[114]

Luther's position on the role of Scripture in cosmology implies that the Christian astronomer was entitled to pursue his own course in the research of nature but that his philosophical presumptions must be compatible with Holy Writ.

> Thus, we Christians should think differently from the philosophers on the causes of these things, and, if some things are over our head (like this matter of the waters above the heavens) it is better to believe, while confessing our ignorance, than either to deny impiously, or to interpret arrogantly on the basis of our comprehension. For it is becoming to preserve the expression of the Holy Scripture and keep attached to the words of the Holy Spirit, whom it has pleased to divide creation in this way—in the middle lay the firmament, produced from the formless heaven and formless earth and given extension by the Word, and then, above and below the firmament, there were waters drawn up from that formless mass. All this the Holy Spirit calls heaven, with seven spheres and the whole airy space, where impressions [phenomena] are produced and in which birds flit about.[115]

112 Ibid., 42: 5: "[...] maneamus discipuli et relinquamus magisterium Spiritui sancto."

113 Ibid., 42: 93 (Commentary on Gen.2:21): "In hos labyrinthos deducitur ratio, cum destitute est verbo et sequitur suum iudicium."

114 Ibid., 42: 96 (Commentary on Gen.2:21).

115 Ibid., 42: 23 (Commentary on Gen.1:6): "Ergo de causis istarum rerum nos Christiani aliter sentire debemus quam Philosophi, etsi quaedam sunt supra captum nostrum (sicut ista hic de aquis supra coelos) ea potius sunt cum nostrae ignorantiae confessione credenda, quam aut impie neganda, aut arroganter pro nostro captu interpretanda. Oportet enim nos servare phrasin scripturae sanctae, et manere in verbis Spiritus sancti, cui ad hunc modum placuit distribuere Creaturas, ut in medio esset Firmamentum productum ex isto informi coelo et terra informi ac per verbum extensum, Deinde supra et infra firmamentum ut essent aquae etiam ex ista informi massa desumptae. Hoc totum Spiritus sanctus appellat coelum cum septem spheris et regione aëris tota, in qua fiunt impressiones, et in qua volucres errant."

Mathieu Arnold remarks that Luther was more prepared to admit his ignorance in his lectures on Genesis than in his sermons.[116] In a theological course of study, there was apparently more room for confessing human ignorance.

Unlike Holy Scripture, the purpose of philosophy was not soteriological but concerned the transmission of knowledge to posterity; it represented knowledge on a human level. Luther argued that philosophers and Christians thought differently about causes.[117] Unlike God's Word, the conjectures of natural philosophy did not represent truth. Luther reminded his readership that although the philosophers determined a fixed number of spheres, this did not mean that things were actually so but that there simply did not seem to be another way of communicating about heaven. If it were not for the imagining of these spheres, not the slightest knowledge about the diversity of planetary motions could be communicated to others. Philosophers taught the foundations of knowledge, Luther stated, and therefore they imagined the celestial spheres:

> The astronomers have stated a fixed number of spheres but this does not mean that things are necessarily so. But it is certain that one cannot communicate the least knowledge of these things if one does not distinguish one sphere from the other according to their diverse motions, and the communication of this, I might say, requires the imagination. It is in this manner that the masters express themselves; "Let us give examples, not because things are this way but because they cannot be taught otherwise."[118]

Moses did not divide heaven into spheres. Commenting on Genesis 1:14, Luther reiterated his statement that the celestial spheres were imagined for educational purposes.[119] To Luther, instruction in philosophical principles was useful for a legitimate practice of the liberal arts and their transmission. He approved of philosophical knowledge to the extent that it lends itself to education.[120] After a brief discussion of the five elements, eight celestial spheres and Aristotle's distinction between the corruptible and incorruptible realms, the

116 Arnold, "De la Création," 22–23.

117 Ibid., 22.

118 Luther, "Enarratio," 42: 21 (Commentary on Gen.1:6): "Sic sphaerarum numerus certus a Mathematicis annotatus est, non quod ita esse necesse sit, sed quod harum rerum aliqualis noticia tradi non potest, nisi sic distinguantur spherae propter diversos motus, ad quos tradendos tali, ut dicam, imaginatione opus est. Sic enim ipsi artifices dicunt: 'Exempla damus, non quod ita sint, sed quod aliter ista doceri non possunt.'"

119 Ibid., 42: 32 (Commentary on Gen.1:14).

120 Ibid., 42: 22 (Commentary on Gen.1:6).

Reformer emphasized that although these things were not certain, it would be barbaric to neglect or disdain them because they comprised the principles of the liberal arts, drawn from the most probable reasons, "especially because in some way they are in agreement with experience."[121] Scripture and the philosophers who study nature have different purposes and speak different languages. On the language aspect, Luther wrote in his comment on Genesis 1:14 that the astronomer and God's Word use different words to designate the same thing: while the astronomer speaks of rays, spheres and epicycles, Holy Scripture ignores these designations and simply uses one word—heaven—to indicate everything that is above us. Time in philosophy means counting motion, whereas to "a Hebrew" it indicates religious feasts and the length of days. Luther added that one art should not mock or condemn another art for its jargon but that the arts instead should help one another.

In concluding this chapter, we might offer some brief remarks finally on Luther's stance towards Copernican astronomy, and the reasons why this will not be taken further into account here. His attitude towards Copernican astronomy is little attested apart from an alleged statement by him made on 4 June 1539, referring to an astrologer who had turned astronomy upside down. Rather than believing this man (even referred to as a *Narr* or fool in one particular version of the *Tischreden*), Luther admonished his listeners to keep to Holy Scripture and believe that the earth stood still instead of the sun, hereby explicitly invoking the Biblical quote regarding the miraculous standstill of the sun effected by Joshua (Josh.10:12: "Then spake Joshua to the Lord in the day when the Lord delivered up the Amorites before the children of Israel, and he said in the sight of Israel, Sun, stand thou still upon Gibeon; and thou, Moon, in the valley of Ajalon."). As a rule, the astrologer in question has been taken to imply Nicolaus Copernicus.[122] Peter Barker and Michel-Pierre Lerner have adduced several counterarguments that question the accuracy of this attribution. The main obstacles concern the identity of the mentioned astrologer and the authenticity of the quote. In addition, the main problem comes with the genre. The genre of the table talks concerns posthumously published compilations of sayings by Luther at dinner conversations (in the refectory of the Augustinian convent in Wittenberg, at his home, or on journeys) which were noted down by others than himself. These compilations were published years after Luther's death. Due to the obscurity surrounding this brief citation,

121 Ibid., 42: 21 (Commentary on Gen.1:6): "[…] praesertim cum aliquo modo cum experientia conveniant."

122 Two among many examples are: Boas Hall, *Scientific Renaissance*, 126; Kraai, "Rheticus' Heliocentric Providence," 94–95, 131.

Luther's position towards a heliocentric astronomy will not be further taken into account.[123]

Luther's metaphysical ignorance, founded on theological anthropology, will be seen to reverberate in the writings of Rheticus, Peucer, Reinhold, and Tycho Brahe. His commentary on Genesis may be regarded as having provided an outline of the rules of rational engagement with visible creation after the Fall in the Lutheran learned context. The next chapter begins with an exploration of how astronomical practice emerged as a form of intellectual devotion.

---

123 For discussions regarding the quote in question, see Barker, "Lutheran Contribution," 34–37; Lerner, "Aux origines de la polémique anticopernicienne (II)," 410–424. The latter compares three different editions in which the quote appeared and provides an extensive discussion of the identity of the astrologer. Both Barker and Lerner suggest that Celio Calcagnini (1479–1541) could be a credible alternative to Copernicus.

CHAPTER 3

Learned Devotion

To the extent that vernacular texts on the natural world in Lutheran Germany argued that the contemplation of the natural world was a source of piety, scientific observation was a constitutive part of Lutheran intellectual piety as well.[1] Deepening the Wittenberg perception of the Lutheran confession as a learned faith, Georg Joachim Rheticus, Caspar Peucer, Erasmus Reinhold, and Tycho Brahe depicted the study of heaven as a means for the individual astronomer to deepen and strengthen their faith. Underlying the idea of astronomical activity as a form of learned devotion was the idea of the divine gift. According to the *Chronicon Carionis*, these gifts were given to the members of the true church to enable them to worship God within His divine creation.[2] Genesis 1:14 accordingly showed that astronomical activity was willed by God. The Wittenberg astronomers saw themselves as putting these divine gifts to use and pursuing the link with the divine. For them, in the words of Miguel Granada, astronomy was "a religious enterprise."[3] By means of contemplating the visible effects of God's creation and providential rule, they were actively engaging in a relationship with God. Their learned activities made them experience and affirm divine creatorship and providential rule of creation. Observing the Book of Nature was an exercise of faith.[4] Crucial in this respect is the understanding of God as an architectonic mind; crucial also was their abhorrence of a world view that was not governed by divine providence, exemplified by the coming-into-existence of the motif of the Epicurean philosopher as an antitype.

1 The Architectonic Mind

From the 1530s onwards, divine providence would become the pre-eminent focus in Melanchthon's natural philosophy, being characterized by a constant emphasis on and repetition of *a posteriori* arguments that confirmed God's

1 Crowther-Heyck, "Wonderful Secrets of Nature," 253–273.

2 *Chronicon Carionis*, 14.

3 Granada, "Tycho Brahe, Caspar Peucer," 565. Granada is arguing for the religious dimension of Peucer's astronomy.

4 Woolford, "Natural Theology," 7.

 DOI:10.1163/9789004387751_005

providence in His creation.[5] The perpetual order in the world proved that God was an architect who maintained this order by His divine providence. Nature had not come into existence by chance but proved instead that the world had originated from an eternal mind.[6] God had left divine signs in nature that were traceable *a posteriori* through experience and reasoning. In tracing these signs, man reinforced the inborn notion of divinity that due to the Fall had become covered over, like mud on "a filthy mirror." This natural knowledge indicated that there was but one craftsman of the world and its order who was wise, benign, just, and true.[7] The study of nature was represented as a journey to acquire knowledge of God, rendered in the term *agnitio Dei*.[8] Sachiko Kusukawa points out that *agnitio* is not be read in its strict meaning as an accumulation of true knowledge. It instead held a Christian connotation, being understood as a recognition of the divine by those who have faith in Christ.[9]

Melanchthon believed the study of nature to be complementary, yet secondary to true doctrine. In combination with the reading of Holy Writ, the study of the testimonies of divinity in nature supported and sustained faith.[10] This supportive role of reason in relation to faith stimulated piety and as such strengthened the church.[11] Although an emphasis on divine providence was not exclusive to Lutheran circles, it featured prominently in Lutheran education throughout the sixteenth century: "a Lutheran must be able to achieve knowledge of God's providential plan" and science was believed to reveal that plan.[12]

Ideas about the architectonic mind resounded throughout the sixteenth century. The first lines of text a student of astronomy was to read in Erasmus Reinhold's commentary on the *Theoricae novae planetarum* of fifteenth-century astronomer Georg von Peurbach were the first stanza of a poem by Melanchthon: "The most beautiful heavenly bodies do not move by coincidence, but are truly governed by the mind and the plan of God."[13] This stanza instantly reminded his students that their subject referred them to divine cre-

5 Kusukawa, "Introduction," xxi–xxii; eadem, *Transformation of Natural Philosophy*, 129.

6 Melanchthon, "Initia," 13: 200.

7 Melanchthon's entry about God in the *Initia physicae doctrinae* provides a full elaboration of this argument; see: Melanchthon, "Initia," 13: 198–202.

8 Ibid., 13: 190.

9 Kusukawa, *Transformation of Natural Philosophy*, 95.

10 Feldhay, "Religion," 3: 737.

11 Methuen, "Role of the Heavens," 393.

12 Barker, "Role of Religion," 61.

13 *Theoricae novae planetarum*, A1v: "Non ferri casu pulcherrima corpora mundi, verum mente regi consilioque dei."

ation. The particular devotional role of mathematical astronomy in relation to faith consisted in affirming the existence of an eternal architect and his providential rule by means of observing the celestial order.[14]

The next step is to depict this intellectual ability as a form of learned devotion, following naturally from the appropriation of the divine gift of the mathematical disciplines. The notion of the gift implies that God had left a divine trace in humankind enabling man to detect the order in the visible world.[15] Of our astronomers under discussion, all four described astronomy, arithmetic, and geometry as divine gifts to humanity, being supernaturally infused into man. Putting these supernatural gifts to use made the astronomer aware of the orderly course of planetary motions; the diligent observer could thus comprehend that heaven was created by an architectonic mind. In the dedication of the Prussian Tables, Erasmus Reinhold portrayed the doctrine of numbers, proportions, shapes (arithmetic and geometry), and celestial motions (astronomy) as "rays of divine wisdom transfused into human minds," being firm testimonies of God and providence and necessary resources for life.[16] Shortly thereafter, he extended his statement.

> And so in order that I may give God the honor He is due, I affirm it to be absolutely true that all this wisdom—the doctrine of numbers, measures and heavenly motions—is a light divinely spread in human minds to show that this world was not concocted by chance from Democritian atoms, but that there is an architect, the eternal, just and benevolent mind and to great benefits to our life.[17]

In the dedication of *Tabulae directionum*, Reinhold again referred to divine light being dispersed in man, meant to enlighten the virtue of his intellect.[18] On a similar note, Caspar Peucer remarked in the dedication to his *Elementa* that the knowledge of numbers was "a trace of divinity impressed upon human minds."[19] The heavenly motions and the celestial bodies' position and shape

14 Barker, "Lutheran Contribution," 44–45.

15 Chassagnette, "La géométrie," 12.

16 Reinhold, *Prutenicae tabulae*, α3^r^.

17 Ibid.: "Itaque ut honorem debitum Deo tribuam, adfirmo id quod verissimum est, totam hanc sapientiam, doctrinam numerorum, mensurarum ac motuum coelestium, lucem esse divinitus sparsam in hominum mentes, ut ostendat hunc mundum non casu ex Democriti Atomis conflatum esse, sed architectatricem esse mentem aeternam, iustam, et beneficam et ut ingentes vitae utilitates adferat."

18 Reinhold *Primus liber Tabulae directionum*, α2^r-v^.

19 Peucer, *Elementa*, 2^v^.

were divine testimonies, and the study of them led to the recognition of the architectonic mind. Again, the recognition of such an architectonic mind implied the certainty that man and the world had not been a result of chance. God had created the visible realm to establish an eternal connection between himself and humankind.[20]

The heavenly motions were presented as enabling a perpetual connection with the divine. Visible creation was conceived of as a gateway to divinity. As a result, appropriating the divinely impressed doctrine of astronomy was introduced by Peucer as a means to (re)affirm and reinforce the link with the divine. This is not limited to astronomy. In *De dimensione terrae* (1550), a work on mathematical geography, Peucer explained that the mathematical study of nature enabled one to discern the traces of divinity left by the creator in the world and in the spirit of man to whom He had given the intellectual capacity to quantify and calculate and, as such, to recognize His work. Peucer used the term *assensio Dei* when referring to this recognition. Arguing for Peucer's understanding of the mathematical disciplines as having a contemplative and devotional function, Axelle Chassagnette explains this recognition as the soul's assent to the existence of God.[21] Like Kusukawa, she does not regard knowledge as the result of this contemplation.

The portrayal of astronomy as learned devotion also surfaces in Rheticus's inaugural lecture as a young professor in lower mathematics:

> For, apart from bringing great benefit to universal life, this most beautiful order of movements also reminds us that nature does not exist by chance, but is created and governed by an eternal mind. Hence, it strongly confirms in the mind worthy beliefs about God and providence. Consequently, among the philosophers only the Epicureans were ungodly, because they did not want to behold these illustrious proofs of God, that is, the most firm laws of movements and this wonderful harmony.[22]

---

20 Ibid., 3^{v}.

21 Chassagnette, "La géométrie," 12–13.

22 "Preface to arithmetic," 94; *In arithmeticen praefatio*, [A6^{r-v}]: "Nam praeterque, quod utilitates magnas affert, ad communem vitam, etiam monet nos hic pulcherrimus ordo motuum, hanc naturam non extitisse casu, sed ab aeterna mente ortam esse et gubernari. Proinde valde confirmat in animis honestas opiniones de Deo, et de providentia. Itaque ex Philosophis soli Epicurei fuerunt αθεοι, quia haec illustria de Deo testimonia, videlicet, certissimas motuum leges, et mirabilem harmoniam aspicere noluerunt."

The final sentence and the use of the Greek word *atheoi* confirm that Rheticus deemed astronomy a devotional enterprise with the capacity to strengthen beliefs about God and providence. In his 1542 oration on astronomy and geography, the mathematics professor linked the ability to detect the architect from the observation of heaven to making appropriate use of the divine gifts:

> As it is most befitting in all things to start with God, so, in this consideration of studies, we should be reminded of the Architect when we contemplate heaven itself. We should consider that He neither established this marvelous order in vain nor handed down knowledge of these movements to humankind without reason. If someone does not think about this, how does he differ from a Cyclops who, spending his life in a cave, neither looks up to heaven nor worships an eternal god? However, if someone, as it is proper for human nature, makes much of God's gifts, he must declare that the laws of these motions were established and the knowledge handed down, for the sake of great benefits.[23]

Homer depicted the Cyclopes as an insolent and lawless race of giants who lived in caves on the island of Sicily, and who, as Rheticus reminded his audience, disregarded Zeus.[24] His imagery of the Cyclops was in itself an affirmation of his appraisal of astronomy as a devotional enterprise, enabling the astronomer to deduce divine presence and providence from the heavenly lights, and to strengthen his religious conviction. A similar statement had been made by Melanchthon some years earlier, in a prefatory letter by him to Reformed theologian Simon Grynaeus (1493–1541) from 1535, inserted in Reinhold's commentary on Peurbach's *Theoricae novae planetarum*.[25]

---

23 "On Astronomy and Geography," 115; Rheticus, "Oratio de astronomia et geographia," A5^r: "Ut autem in omnibus rebus à Deo ordiri rectissimum est, ita in hac quasi deliberatione de studiis, cum caelum ipsum contemplabimur, veniat in mentem Architecti. Hunc cogitemus: nec frustra hunc mirandum ordinem instituisse, nec frustra generi humano noticiam horum motuum tradidisse. Haec si quis non cogitat, quantum à Cyclope illo distat, qui in antro degens, nec caelum aspicit, nec numen ullum aeternum veretur? Si quis autem ut hominis naturae convenit, Dei munera magni facit, is statuet magnarum utilitatum causa has motuum leges institutas, et doctrinam traditam esse."

24 Schmitz, "Cyclo'pes," 1: 909.

25 Melanchthon, "Doctissimo viro Simoni Grynaeo," B3^r. Melanchthon's letter to Simon Grynaeus was an encouragement for students to study astronomy and cosmology, and it was republished many times in conjunction with astronomical works internationally and transconfessionally; see: Valleriani, Federau and Nicolaeva, "Hidden Praeceptor," 409.

The notion of God as the architect of the world is strongly reminiscent of the age-old Platonic description of God as a geometer, implying that God had ordered the entire universe in line with a divinely conceived geometrical plan.[26] Melanchthon equally linked the metaphor of God as a geometer to His providential rule of the world. Only by studying the world, in agreement with God's will, would the divine traces become visible:

> On the contrary, just as Plato said [Plutarch, Conviv disp 8.2] that God always geometrizes—that is, governs these inferior matters by measuring all things with a very certain motion—so let us, in considering the lines of this superb master, take pleasure in this wonderful Geometry that shows divinity to us. It is for this reason above all that celestial nature was created, to show us God, and so it suits the will of God not to be visible to those who neither behold nor inquire into these traces of divinity.[27]

In these prefaces and orations, the astronomers promised their readership the affirmation of God as the architectonic mind responsible for creating the world.

Tycho Brahe's inaugural lecture at Copenhagen University, delivered in 1574, shows that the notions of the architectonic mind and the visible creation as a testimony of divine providence continued well into the sixteenth century and well outside the homeland of the Reformation. The Danish astronomer argued that the observation of heaven demanded the recognition of divine creatorship and divine providence: the visible creation of heaven was a place where God could be admired and glorified.[28] Moreover, it was proper to human nature to consider heaven as such. The glorification of God was advocated as a theological justification for observation.[29] Tycho claimed that the eternal laws of celestial motion and harmony in heaven taught humans about the "chief cause that is outside the visible part," and that, "following the true and coinciding knowledge of God that has been revealed to us in the Word given by Him." Nothing was more in agreement with man's nature, his end and his position on the earth—the center of the universe—, than to recognize God "in His very

26 Barker, "Role of Religion," 82; Harrison, *Fall of Man*, 102.

27 Melanchthon, "Doctissimo viro Simoni," B3v: "Quin potius ut Plato dixit deum semper γεωμετρειν, hoc est, certissimo motu omnia metientem, gubernare hac inferiora, ita nos vicissim huius summi artificis lineas considerantes, hac pulcherrima geometria nos oblectemus, quae divinitatem nobis ostendit. Si ob hanc causam praecipue condita est coelestis natura, ut certe est, ut Deum nobis monstret, satis constat voluntati Dei non parere istos, qui haec divinitatis vestigia non aspiciunt neque inquirunt."

28 Brahe, "Oratio de disciplinis mathematicis," 1: 152.

29 Methuen, "This Comet," 99.

wise and various works," and to "glorify Him with dignity and praise."[30] God's "wisdom and elevation shine nowhere more than in the immense and eternal theater of the heaven and the stars."[31] Nature served, in other words, to glorify and admire God. Tycho upheld the Lutheran notion that only aided reason could reach this spiritual perspective on the world. He concluded the paragraph with a quotation from Helius Eobanus Hessus (1488–1540), a German poet from the region of Hessen-Kassel: "He who denies that God exists, must look up to the heaven and the stars: He who contemplates the stars, swears that God exists."[32]

Observation was a source of piety. Tycho depicted heaven as an intermediate realm where divine providence produced visible effects. Based on Genesis 1:14, he argued that the observation of heavenly phenomena enabled man to distinguish time, days, and years and to notice divinely created signs in heaven.[33] The Danish astronomer shared Melanchthon's view of the world as a place ruled by divine providence.[34] This means that the world was not just divinely created but was in fact the ongoing result of divine creation: God did not cease to rule the world or to intervene after its creation: the celestial effects were the continuous effects of His creatorship and providence.

Clearly, the stars had not lost their status as objects of religious contemplation. Katherine Park, Gianna Pomata, and Lorraine Daston seem to imply that, along with the evolution of the early modern rise of observation as a cognitive category—in which Tycho played a central role—, the act of observing no

30 Brahe, "Oratio de disciplinis mathematicis," 1: 152: "Quorsum vero sapiens ille et providus universitatis autor, tam admirandas et perpetuas motuum coelestium leges, tanta diversitate, et tam concinna harmonia effinxisset, si eas ab hominibus, quorum etiam causa magna ex parte visibilia condidit, ignoraro voluisset? Imo potius haec indefesso labore perscrutari ac inquiri voluit, ut ipsius majestas et sapientiae quoque, hinc a mortalibus perspici et celebrari possit. Quapropter post Dei veram et competentem cognitionem, nobis in verbo a se dato revelatam, nihil hominis naturae magis proprium, et fini, propter quem homo in Terra, mundi centro, conditus et collocatus sit, magis consentaneum esse judico, quam ut inde tanquam e loco medio, ea, quae in tota mundi fabrica, imprimis vero in coelesti illa et fulgentissima tot perpetuarum stellarum regia elucent, prospectans, jucunda hac et ingeniossa contemplatione aetatem suam suaviter transigat, Deumque opificem in his suis sapientissimis et variis operibus agnoscens, merita eum veneratione ac laude celebret."

31 Ibid., 1: 152: "Quamvis enim in hac inferiori natura mirabilia extent, divinae eius sapientiae testimonia: tamen nusquam magis elucet eius sapientia et majestas, quam in perpetuo illo et immenso coeli stellarumque omnium theatro."

32 Ibid., 1: 152: "[...] unde non immerito dixit Eobanus: Qui negat esse Deum, coelos et sidera spectet: Sidera qui spectat, jurat is esse Deum."

33 Ibid., 1: 154.

34 Mosley, *Bearing the Heavens*, 158.

longer implied a spiritual dimension, or at least they fail to mention it as doing so.[35] Tycho's case shows precisely the reverse. Reputed as a pioneer of early modern scientific observations, his writings disclose that the religious significance attributed to astronomical observation did not diminish in relation to the increasing understanding of scientific observation as a cognitive category. Quite the contrary, putting divine gifts into practice by observing heaven and thereby inciting piety can be described as a form of learned devotion.

2 The Antitype of the Epicurean Philosopher

Almost as a rule, references to the architectonic mind became paired with allusions to antitypes who failed to acknowledge the presence of divine traces in the world. A first such analogy was the godless Cyclops; a more frequent example in this regard is the Epicurean philosopher. The historiographical theme of Epicurean philosophy has been little documented for the sixteenth century. Still, based on Catherine Wilson's account of Epicureanism in early modern philosophy, it is noteworthy that the Epicureans were apparently depicted by Melanchthon, Luther, and the Wittenberg astronomers in the same manner as they were in the seventeenth century: "Epicureanism was almost synonymous, for early modern philosophers, with atheism, and so with the denial of divine creation, providence, the immortality of the soul, heaven and hell, and moral right and wrong."[36] Nonetheless, this account conflicts with the real Epicurus who had admitted the existence of the gods, insisting that piety was to be cultivated.[37] Alongside Wilson, Christoph Lüthy noticed the same discrepancy between the real Democritus and the fictional images that were going around in the seventeenth century.[38] The following paragraph reconstructs a fictional image of the Epicurean that was dominant at least in Wittenberg University, and by extension most likely in Lutheran circles.

The labels of Epicurean and Democritian seem to have been used interchangeably. Whereas Peucer and Reinhold referred to the Democritian atoms, Melanchthon spoke of Epicurean atoms.[39] Compared with the theory of Democritian atoms, the Epicureans were mentioned much more often—hence the

35 Park, "Observation in the Margins"; Daston, "Empire of Observation"; Pomata, "Observation Rising."
36 Wilson, "Epicureanism in Early Modern Philosophy," 267.
37 Ibid.
38 Lüthy, "Fourfold Democritus," 443.
39 Melanchthon, "Initia," 13: 191.

present discussion of the Epicurean instead of the Democritian philosopher. The main source of dissatisfaction on the part of the Wittenberg side was the Epicurean denial of a world that was governed by causality. In the specific context of Wittenberg University, the ancient Epicurean philosopher emerged as the ungodly antitype of the faithful Wittenberg astronomer. Patrick Boner contends that the term Epicurean was being used as a general appellation for any advocate of non-causal explanation. In reference to Johannes Kepler's discussions of the new star that had arisen in 1604, Boner found the astronomer railing against the Epicurean erosion of causality in favor of coincidence. The new star had not come about by chance but was the product of an architectonic mind.[40] Likewise, in the eyes of Melanchthon and the Wittenberg astronomers, the Epicureans failed to recognize God as the final and efficient cause of creation.[41] By contrast, divine creatorship and providence precisely were the fundamentals of Melanchthon's natural philosophy. Hence, the combined argument of affirming the existence of an architectonic mind by means of celestial effects and denial of a coincidental universe proves to be a recurrent theme in early modern Lutheran astronomical writings. Due to the Epicurean denial of divine creatorship and providence, Epicureanism became linked with godlessness.[42] In a Lutheran context, this ungodliness was explained as the result of being detached from faith or from God's Word. In his commentary on Genesis, Luther himself considered the Epicurean opinions as the result of human reason deprived of God's Word, gone astray in its own labyrinth and following its own judgment.[43] The Epicureans, so to speak, were pictured as proceeding in the obscurity of the unfaithful mind. In the textbook *Erotemata dialectices*, Melanchthon equally rebutted the "Epicurean ravings" for teaching things incompatible with true doctrine.[44] In the eternal academy, he wrote, man will be glad to have defended what is true and pleasing to God and to have renounced the ravings of the Epicureans.[45]

Examples of the link between Epicureanism and godlessness are found in astronomical contexts as well. Rheticus reproached the Epicurean philosophers for their denial that the stars were true works of God. Assuming the sun to be a vapor set alight by its fall and carried by its fall—and thus, excluding divine intervention—was an "insult" against God. One must refrain from such impi-

40 Boner, "Kepler v. the Epicureans."
41 Melanchthon, "Erotemata dialectices," 13: 656–657.
42 Barker, "Lutheran Contribution," 45.
43 Luther, "Enarratio," 42: 93 (Commentary on Gen.2:21).
44 Melanchthon, "Erotemata dialectices," 13: 658: "Epicurea plena est horribilium furorum."
45 Melanchthon, "Hier. Baumgartnero," 3: 914.

eties that are opposed to God's Word.[46] The notion of looking upon the celestial motions in an ungodly fashion had yet another implication for the Wittenberg professors: it implied the denial of the immortality of the soul. In a prefatory letter to Simon Grynaeus in an edition of Johannes de Sacrobosco's *De sphaera* (1531), Melanchthon advanced the idea that men were given eyes for the sake of astronomy, explicitly drawing on Plato (*Republic* VII, 530d, and *Cratylus* 399b). Melanchthon later buttressed his view that observation of celestial order effected the observer's rational acknowledgment of God as the creator of the celestial phenomena:

> For it is not possible for the human mind not to conclude that there is a mind that rules and governs everything, if it contemplated these established courses and laws of the great circuits and stars. For no such thing can exist or continue by chance or by another power without a mind. For this reason—if astronomy corroborates the belief about God in the minds of men—we have to consider that Plato said not only learnedly but also piously that eyes are given to use because of astronomy. [...] Furthermore, only those who spurned astronomy were professedly ungodly having done away with providence, they also removed the immortality of our souls. If they had reached this knowledge [that eyes were given to men for the sake of astronomy], they would have perceived the manifest traces of God in nature, and, having noticed them, they would have been forced to acknowledge that the universe is made and governed by a mind.[47]

The similarity in wording in comparison with the above quotes of the astronomers strongly suggests that Melanchthon must have had in mind the Epicurean philosophers or the antique theory of Democritian atoms when speaking of "those who spurned astronomy." His words show that the recognition of God

---

46 *In arithmeticen praefatio*, [A6^r].

47 Melanchthon, "Preface to *On the Sphere*," 106; Melanchthon, "Doctissimo viro Simoni Grynaeo," 2: 531: "Neque enim fieri potest, quin statuat humanus animus, mentem aliquam esse, quae regat et gubernat omnia, si animadverterit illos ratos cursus et leges magnorum orbium et stellarum. Nihil enim tale casu aut alia vi ulla sine mente existere aut constare potest. Quare si astronomia confirmat hanc de Deo opinionem in animis hominum, Plato non solum erudite, sed etiam religiose dixisse iudicandus est, astronomiae causa nobis oculos datos esse. [...] Proinde ex philosophis soli isti, qui astronomiam aspernati sunt, ex professo fuerunt αθεοι, et sublata providentia etiam immortalitatem animorum nostrorum sustulerunt: qui, si attigissent hanc doctrinam, manifesta Dei vestigia in natura deprehendissent, quibus animadversis coacti fuissent fateri, mente aliqua hanc rerum universitatem et conditam esse et gubernari."

as the eternal architectonic mind—achieved by means of visual observation of the divine traces in nature—implied the acknowledgment of divine providence which, in turn, implied a recognition of the immortality of the soul. In addition, the Epicurean outlook on a world created by chance precluded their seeing traces of divinity in celestial revolutions, and precluded the possibility of their achieving *agnitio Dei*.

> As for these Epicureans, who are neither astonished at the wonderful motions of the celestial bodies, nor contend that the knowledge of them is useful, I judge them unworthy of being called men. For truly, they not only wage war against human nature, which is created in particular to look upon these divine things, but they even fight against God. After all, God wanted the awareness of these marvelous revolutions and virtues to lead us to the knowledge of divinity.[48]

In sum, the Epicurean philosopher was truly an antitype to the self-representation of Melanchthon, Peucer, Rheticus, and Reinhold. In the eyes of the faithful scholars at Wittenberg University, the Epicurean denial of a divinely established world was anathema.

48 Melanchthon, "Doctissimo viro Simoni Grynaeo," B3^{r}: "Nam Epicureos illos, qui neque pulcherrimos motus coelestium corporum admirantur, neque cognitionem eorum utilem esse contendunt, ne quidem appellatione hominum dignos esse iudico. Etenim non solum bellum gerunt cum humana natura, quae praecipue ad has divinas res aspiciendas condita est, sed etiam θεομαχοι sunt. Voluit enim Deus horum mirabilium cursuum ac virium noticiam, ducem nobis esse ad divintatis cognitionem."

CHAPTER 4

Georg Joachim Rheticus and Fruitful Ignorance

The cosmological writings of Georg Joachim Rheticus, Caspar Peucer, and Tycho Brahe reveal a deep motivation to theorize piously about the heavens, unsurprising for members of the true church whose responsibility it was to keep philosophy—being the handmaiden of theological debate—pious. Martin Luther's pessimist epistemology build on reason's postlapsarian condition of metaphysical ignorance and its tendency to err. This epistemology clearly resonated in the Lutheran astronomers' vindications of non-canonical cosmological claims, be it their own or those of others. Epistemic pessimism was constitutive of Lutheran intellectual piety in general, and of the freedom to philosophize propounded by these astronomers in particular. A Lutheran pessimist epistemology has also been traced in the works of Philipp Melanchthon. Sachiko Kusukawa demonstrated convincingly that the *praeceptor Germaniae* transformed Scholastic natural philosophy into a Lutheran variant, precisely by modeling it to the principle of the mistrust of human reason, which was a major departure from Scholastic thinking.[1] Even though these astronomers were educated within the university system and/or were professionally active at universities, their liberty to philosophize manifested itself in correspondence and treatises—whether or not self-published—in the extra-academic context of, for example, a grand tour, a court or an island that harbored an astronomical observatory. The anti-authoritarian argument they advocated was articulated in theological anthropological terms; it was motivated by a humble awareness of man's sinful condition after the Fall. At the Academia Wittebergensis by contrast cosmological discussions will be seen to have been settled in favor of the canon. The choice to proceed by authority was motivated by the same theological anthropological concern of intellectual piety mentioned when advocating philosophical liberty.

Copernicus's physical hypothesis was instructed to young students at Melanchthonian universities only at the turn of the 1590s. Reasons for the delayed introduction have recently been found in the humanist pedagogical program at Wittenberg University. Omodeo and Regier point to Melanchthon's perception of Copernicus's model as a threat to the order of the liberal arts, and

1 Kusukawa, *Transformation of Natural Philosophy*. This view was supported in Cabezón, "Introduction," 5; and Peterson, "Philipp Melanchthon," 69–75.

 | DOI:10.1163/9789004387751_006

thus to the institution of the university itself.[2] According to Stefan Kirschner and Andreas Kühne, the decline of medieval Scholastic disputation culture at universities was one of the reasons that Copernicus's cosmological theses were not the subject of extensive university debate in the second half of the sixteenth century. Wittenberg University too witnessed a decay of the disputation culture at the end of the second decade of the sixteenth century due to the humanist critique of Scholastic learning, the rise of humanist genres such as the declamation, and the more practical needs of university education that were not met by intellectual debate. The humanist literary forms of the textbook, the treatise, and the oration that became increasingly popular during the Renaissance, implied a new form of instruction. The textbook presented basic and commonly accepted views.[3] The textbook, the university, and the professor represented authoritative knowledge.[4] The genre of the textbook fitted the purpose of academic instruction and was apt not to sow any of the kinds of doubts by which the mind could have been further corrupted.

1 The Renaissance Restoration of Astronomy

Historiography has assessed the precise meaning of the Renaissance "restoration" of astronomy in various manners. Marie Boas Hall indicated that there was more to the renovation of astronomy than the late-fifteenth century recovery and translation of ancient mathematical texts. A profound study of Ptolemy's *Almagest* was but the first step in the process of astronomical reform. In reference to the work of Tycho Brahe, other appraisals of this renovation have surfaced.[5] Ann Blair, for instance, argues that the term restoration "rather than a return to any particular cosmological system or set of parameters, referred more loosely to an unspecified ideal astronomy from a mythical past, perfect in all respects."[6] Adam Mosley, on the other hand, adds to the recovery of ancient texts the search for a universally valid astronomy based on the belief in the stability and the regularity of heavenly bodies and motions.[7]

2 Omodeo and Regier, "Wittenberg Reception of Copernicus," 87–88.
3 Kirschner and Kühne, "Decline," 13–41.
4 Kirwan, "Introduction," 5.
5 Boas Hall, *Scientific Renaissance*, 125–127.
6 Blair, "Tycho Brahe's Critique," 357.
7 Mosley, *Bearing the Heavens*, 59.

The sixteenth-century astronomical revolution in which the Lutheran-Wittenbergian astronomers under discussion took part represented, on the one hand, the disciplinary incursions of the astronomers into the field of cosmology, and on the other, the departure from the tradition-bound way of Scholastic thinking with a strong focus on the opinions of Aristotle.[8] A crack in the Scholastic hierarchy of sciences had already come to the surface at Padua University in the fifteenth century, well before Copernicus's introduction of a heliocentric model of the universe. Here the mathematician Johannes Müller von Königsberg, *alias* Regiomontanus, delivered an inaugural lecture in 1464. Regiomontanus, who was praised as a restaurator of the art of astronomy by his sixteenth-century fellow astronomers, was the best-known early proponent of the Renaissance search for a renovation or restoration of astronomy.[9] His legacy and fame continued well into the sixteenth century, as demonstrated, for example, by Reinhold's declamation on Regiomontanus at Wittenberg University. Out of a feeling of dissatisfaction with the contemporary pursuit of astronomy, a concern for disciplinary renewal had arisen in the fifteenth century. In 1464, the German mathematician proclaimed that the mathematical disciplines were characterized by a "certitude" which the other arts did not possess, and argued for the necessity of the quadrivium in philosophy for a true understanding of philosophers' books.[10] His elevation of the mathematical disciplines was paired with, as Noel Swerdlow puts it, "a complete repudiation of Scholastic natural philosophy and its pretensions to knowledge," which is a preview of the reproaches made by Rheticus and Tycho Brahe.[11] Regiomontanus criticized the Scholastics for neglecting the quadrivium and for interpreting Aristotle in a selective and haphazard manner. While the philosophy of Aristotle was modified by centuries of human interpretation, Regiomontanus stated that Euclid's theorems were timeless and unchanging: their certainty remained, whether read today or a thousand years ago. The discoveries by Archimedes "will instill no less admiration in men to come after a thousand centuries than the delight instilled by our own reading."[12] So far, his lecture is considered the earliest example of a repudiation of Scholastic philosophy based on an unfavorable comparison of mathematical certainty and philosophical uncertainty. According to Paul Rose, the contrast between the

8 Cabezón, "Introduction,"4.

9 Rose, *Italian Renaissance of Mathematics*, 90.

10 "Oratio Iohannis de Monteregio," C4^r^.

11 Swerdlow, "Science and Humanism," 149.

12 "Oratio Iohannis de Monteregio," C4^v^, C5^r^: "Inventa Archimedis post mille secula venturis hominibus non minorem indecent admirationem, quam legentibus nobis iucunditatem."

immutable certainty of mathematics and the perpetual disputes of philosophy characterized the whole of Regiomontanus's thought.[13]

The Scholastic assessment of natural philosophy as a higher discipline was based on its subject matter, namely the nature of things. It was "a qualitative science of the natural world that explained why things happen in terms of the essential natures of bodies."[14] Due to the unquantifiable nature of the invisible causes (why things happen), natural philosophy could yield probable but not certain accounts. Aristotle's positions on the causal accounts of nature were fully accepted on grounds of the authority of the Stagirite. In general, Scholastic thinking was committed to the preservation of tradition.[15] The authoritative weight of probable opinions was determined by the source from which the opinion was extracted. The more authoritative the source, the more probable the trustworthiness of the statement.[16] "Probability" in this context had the connotation of "worthy of approval."[17] Approvable opinions were assumed to be found in authoritative texts of ancient authorities such as Aristotle and Ptolemy. Absolute certainty, however, was the province of demonstration. True knowledge or *scientia* was reserved for the mathematical disciplines, for they were capable of yielding absolute certainty based on demonstrations of visible, quantifiable characteristics.[18]

Regiomontanus's advocacy of mathematical certainty to the detriment of philosophical uncertainty was to gain momentum in the sixteenth-century astronomical revolution. Rheticus, Brahe, and Peucer would reframe this disciplinary affair in Lutheran anthropological terms.

2 Rheticus's Advocacy of Heliocentrism in Historiography

Georg Joachim Rheticus believed that the cosmological hypotheses of his teacher Nicolaus Copernicus would effect a restoration of astronomy. Born Georg Joachim Iserin, the young Wittenberg professor was the only pupil of the Polish astronomer, at least so far as we know.[19] His visit to Copernicus in Frombork (Frauenburg), a cathedral city in the province of Royal Prussia, then

---

13 Rose, *Italian Renaissance of Mathematics*, 97.
14 Dear, *Discipline*, 3.
15 Cabezón, "Introduction," 4.
16 Dear, "Totius," 146.
17 Hacking, *Emergence of Probability*, 18.
18 Jardine, "Epistemology of the Sciences," 693.
19 For an extensive biography of Georg Joachim Rheticus, see: Burmeister, *Georg Joachim Rhetikus*, vol. 1; Rosen, "Rheticus, Georg Joachim," 11: 395–398.

part of the historical Kingdom of Poland (1385–1569), was part of a tour across several astronomical centers that he made as a young mathematics professor. This tour started in 1538 and temporarily interrupted his professorship at Wittenberg, where he had been appointed to the chair of lower mathematics in 1536. On the advice of German mathematician Johannes Schöner (1477–1547), whom he had met in the largely (politically) independent territory of Nuremberg, he visited Copernicus in 1539, and would remain with him for two years. It was Rheticus who persuaded his teacher to have *De revolutionibus orbium coelestium* published. It was Rheticus who took the manuscript to the office of Nuremberg printer Johannes Petreius. Only in 1541 did he return to Wittenberg, when he was elected dean of the Arts Faculty for the winter semester. Afterwards, Rheticus took academic leave to supervise the printing process of *De revolutionibus* until he departed for Leipzig in October 1542 where he was offered a position as mathematics professor.[20] Andreas Osiander (1498–1552), a Lutheran theologian, took over the supervision of the publication project, and added a notorious yet anonymous preface to it. Rheticus would never return to Wittenberg, although he did remain in friendly contact with Caspar Peucer and Philipp Melanchthon.[21] In the early 1550s, he studied medicine at the University of Prague. Afterwards, he moved to Cracow and acted as a practicing physician there for about twenty years during which he did not cease his mathematical activities.[22] He died in 1574.

Rheticus played a decisive role in the early dissemination of Copernicus's heliocentric astronomy. Not only did he secure the printing of *De revolutionibus*, he also introduced the work of the Polish astronomer at the universities of Wittenberg and Leipzig, and presented the world with two treatises written in defense of the legitimacy of Copernicus's physical claims around 1540. The *Narratio prima* presented a non-technical outline or first account of Copernicus's hypotheses in the form of a letter addressed to Johannes Schöner, published in 1540 in anticipation of the publication of *De revolutionibus*.[23] The *Narratio prima* marks the beginning of the broader reception of Copernicus's heliocentrism. The outline of the Copernican heliocentrism in chapters seven to eleven was of especially significant pedagogical value.[24] The second treatise,

---

20 Gingerich, *Annotated Census*, xii.

21 Burmeister, *Georg Joachim Rhetikus*, 1: 12, 34, 103, 121. One proof of the friendly contact was Melanchthon's invitation of Rheticus to the wedding of Caspar Peucer and Melanchthon's daughter which took place in 1550.

22 Tredwell and Barker, "Copernicus' First Friends," 144.

23 Gingerich, *Annotated Census*, XII; Rheticus, *Ad clarissimum virum ... Narratio prima* (hereafter: *Narratio prima* or First Account).

24 Lerner, Segonds, and Verdet, *Nicolas Copernic*, 1: 421.

the anonymous and undated *Epistola de terrae motu* or Letter on the Motion of the Earth, argued for the compatibility of Copernicus's hypotheses with Holy Writ.[25] If these two treatises can be seen to function jointly as a diptych of Rheticus's defense of Copernican astronomy—with the *Narratio prima* considering the validity of the disciplinary incursion into cosmology by a theorizing astronomer, and the *Epistola* testing the hypotheses against Holy Writ—the opening sentence of the latter functions as a hinge between the two:

> It is mathematically proven that the mobility of the earth is to be assumed in order to obtain a lasting explanation of the phenomena of the heavenly bodies. What, however, is said on this subject according to Holy Scripture?[26]

The usual historiographical portrait of Rheticus is characterized by two key elements: certainty and exceptionalism. As a rule, the Wittenberg mathematician is portrayed as one of the handful sixteenth-century astronomers to whom Copernicus's astronomical hypotheses were cosmological reality. Based on their recent reading of the *Narratio prima*, Michel-Pierre Lerner, Alain-Philippe

25 "Epistola de terrae motu." [1542?], ed. and transl. Reijer Hooykaas. The Latin text in Hooykaas's edition runs from pages 43 to 64; Hooykaas's English translation from pages 65 to 101. "*Epistola de terrae motu*" is the inscription on the title page. "*Diss*[*ertatio*] *de Hypoth*[*esibus*] *Astron*[*omiae*] *Cop*[*ernicanae*]" is the heading. How to designate the *Epistola* is a slightly contentious issue because neither the heading nor the inscription on the title page were decisions made by the author himself, but were instead made by the seventeenth-century printer Johannes van Waesberge. Robert Westman uses as a title "Opusculum (quo a sacrarum scripturarum dissidentia telluris motus emonstrat)," based on an external reference to the "little work" written by Rheticus that appeared in a letter of Tiedemann Giese to Rheticus. Kenneth Howell uses the designation "Treatise on Holy Scripture." Jesse Kraai uses "Treatise on Holy Scripture and the Motion of the Earth." The present book advances the title "Epistola de terrae motu" or "Letter on the Motion of the Earth" in analogy with the inscription on the title page. For reasons of convenience, the footnote references to citations that are rendered in the text will indicate the paragraph number assigned to it by Hooykaas between brackets. References to the Latin text will be headed by the designation [Rheticus], "Epistola"; to Hooykaas's English translation as Hooykaas, "The Anynomous Treatise," and to his personal reflections and commentary as Hooykaas, *G.J. Rheticus' Treatise*. In 2015, a German translation was published in "Die Abhandlung De Terrae motu von Georg Joachim Rheticus." In Kirschner, Kühne, and Schmeidler, *Receptio Copernicana*, 69–90; see: Howell, *God's Two Books*, 58; Kraai, "Rheticus' Heliocentric Providence," 131; Westman, *Copernican Question*, 13–131.

26 "Epistola," 43 (1): "Mathematice demonstrator, ad perpetuam rationem των φαινομενων corporum coelestium obtinendam, terrae mobilitatem assumendam. Quid autem hac de re secundum Sacram Scripturam statuendum est?"

Segonds, and Jean-Pierre Verdet also advanced the idea that Rheticus thought of Copernicus's hypotheses as a depiction of the real *forma mundi*.[27] In 1975, Robert Westman made a highly influential distinction in this regard between Rheticus and his Wittenberg colleagues. Around Melanchthon and most probably Reinhold, a methodological consensus had developed among the Wittenberg astronomers who adopted Copernicus's mathematical advancements (for instance, the ability to calculate planetary positions without using equants) but ignored the cosmological implications of the heliocentric model such as the relative ordering of the planets. Erasmus Reinhold, for example, did calculate his *Tabulae Prutenicae* according to Copernicus's hypotheses but remained silent about their physical status.[28] Rheticus however was an outsider to what Westman coined "the Wittenberg Interpretation" or Nicholas Jardine described as "the pragmatic compromise."[29] Nonetheless, Reinhold's astronomical tables, published in 1551, did prove an important vehicle, however, in the spread and circulation of the Copernican mode of calculus.[30]

Elaborating the notion of the physical reality of the heliocentric system, Katherine Tredwell and Peter Barker termed ten sixteenth-century astronomers "physical Copernicans."[31] Rheticus's exceptionalism was further reinforced by psychological portraits suggesting an odd personality and a peculiar relationship to Copernicus.[32] At the turn of the twenty-first century, Peter Barker and Bernard Goldstein framed the difference between the Copernican astronomers and their Wittenberg counterparts (apart from Rheticus) in terms of Aristotelian logic. Barker and Goldstein argue that Copernican astronomers considered the heliocentric model as the *propter quid* of the celestial effects

---

27 Lerner, Segonds, and Verdet, *Nicolas Copernic*, 1: 422.

28 Ibid., 1: 399.

29 Westman, "Melanchthon Circle," 166–167: Jardine, *Birth of History and Philosophy of Science*, 293–243. A similar observation has recently been made for seventeenth-century Jesuit science in response to the publishing of the *Almagestum novum* by Giovanni Battista Riccioli (1598–1671) in 1651, namely, that when presented as a computational device, heliocentric theory was tolerated; see: Gambaro, "Geo-heliocentric models," 265–294.

30 Lerner, Segonds, and Verdet, *Nicolas Copernic*, 1: 400. For further information on the importance of the *Tabulae Prutenicae* in the early phase of the reception of Copernican heliocentrism, see idem, 1: 403–409.

31 Tredwell and Barker, "Copernicus' First Friends," 143–166.

32 An example is the psychodynamic hypothesis developed by Robert Westman. Rheticus's relation to Copernicus and older men in general is said to have been profoundly influenced by the beheading of his own father Georg Iserin when he was a young boy, and by the subsequent consequences of this event for his family life. His admiration for Copernicus was founded on his perception of the astronomer as a kind and strong father: see Westman, "Melanchthon Circle," 185–190; idem, *Copernican Question*, 148–150.

whereas the non-Copernicans found themselves restricted to delivering *quia* demonstrations. To the latter group, the astronomer's earthbound position was a practical hindrance in achieving true causal knowledge of the phenomena. Despite its practical impossibility, the *propter quid* demonstration remained the cognitive ideal. Hence, Barker and Goldstein designated the Wittenberg astronomers as "perpetually frustrated realists."[33] Whereas the Copernicans approached astronomy as an *a priori* science, the non-Copernicans thought of their art as *a posteriori*.[34] Rheticus being a Copernican astronomer should have believed that the heliocentric model was the *propter quid* of the celestial phenomena, and considered astronomy to be an *a priori* science. Unlike his fellow Wittenberg astronomers, Rheticus held no epistemological reservations about the knowability of the world. Copernicus's heliocentric universe was ontological truth. On a same note, Reijer Hooykaas, Robert Westman, and Kenneth Howell saw no point to a Scriptural defense of the motion of the earth by Rheticus such as the *Epistola de terrae motu*, if not to defend a physically realist position.[35] The supposed physical realism of other sixteenth-century astronomers, such as the Louvain mathematician Gemma Frisius (1508–1555) and even Copernicus himself, has been recently questioned as well. According to a number of scholars, Copernicus knew he lacked conclusive proof for the heliocentric model to be a true representation of the physical structure of the universe; in terms of certainty, his hypotheses rendered only a probable and fallible account of the universe.[36] Rheticus's physical realism has so far gone unchallenged, apart from a cautious suggestion by Tredwell and Barker in 2004. Based on a citation from the *Narratio prima*, they suggest that Rheticus might

33 According to the Aristotelian ideal of scientific knowledge, there are two patterns of syllogistic demonstration; the *propter quid* pattern comprises reasoning from the true cause to effects (*a priori* demonstration) and the *quia* pattern reasoning from effects to possible causes (*a posteriori* demonstration). An ideal pattern of demonstration, called the demonstrative regress (*regressus demonstrativus*), first draws up a multiple number of possible *quia* demonstrations (in this case hypotheses) that explain the observed effect (in this case the celestial phenomena), then selects the true cause among them and finally concludes with a *propter quid* demonstration that shows how this true cause renders the original effect; see: Barker and Goldstein, "Realism," 232–258.

34 Barker, "Role of Religion," 71–81.

35 Hooykaas, *G.J. Rheticus' Treatise*, 102, 107; Howell, *God's Two Books*, 58, 61; Westman, *Copernican Question*, 131. According to Westman, the *Epistola* represents a joint effort by Rheticus and Copernicus who worked out the basic elements of a systematic defense of the compatibility of Scripture with the new hypotheses. For now, at least, there is not sufficient proof to substantiate this thesis.

36 Hallyn, "Gemma Frisius," 83; Swerdlow and Neugebauer, *Mathematical Astronomy*, 1:26–27; Westman, "Astronomer's Role," 112.

not have been entirely convinced that Copernicus proved the motion of the earth beyond doubt, but the inconclusive evidence led them to emphasize the speculative nature of his suggestion.[37]

The incentives to start questioning Rheticus's physical realism were the latter's skeptical remarks in the *Epistola de terrae motu*, hinting at the human inclination to be deceived in thinking about that world and a limited knowability of the visible world. Reijer Hooykaas was one of the few who has considered these skeptical remarks but assumed that their presence derived from mere rhetorical-strategic concerns.[38] Recently, Michel-Pierre Lerner, Alain-Philippe Segonds, and Jean-Pierre Verdet have deemed these skeptical statements to be of a diplomatic nature because the theme of the limits of human knowledge was dear to theologians.[39] When taken seriously, however, the anonymous treatise tells a pessimist narrative of human knowledge after the Fall. So, historiographical procedures will be reversed: instead of tailoring the reading of the *Epistola* to the physical Copernicanism supposedly displayed in the *Narratio prima*, a theological anthropological reading of the *Epistola* serves as a point of departure to understand Rheticus's support of Copernicus's disciplinary role-reversal. His exceptionalism will be diluted to the point that his writings on Copernican astronomy can readily be reintegrated within the Wittenberg and Lutheran religious and theological frame of reference. The following thus constitutes an alternative reading to Hooykaas's interpretation of the *Epistola*.

3 *Epistola de terrae motu*: A Theological Anthropological Turn

For centuries, the only indication that there existed something like a Scriptural defense of the earth's motion written by Rheticus was a reference by bishop of Chelmno Tiedemann Giese (1480–1550), whom Copernicus identified as one of the friends who had persuaded him to publish his work.[40] In a letter dated 26 July 1543, Giese wrote to Rheticus about a little work

37 Tredwell and Barker, "Copernicus' First Friends," 145–146.

38 Hooykaas, *G.J. Rheticus' Treatise*, 117.

39 Lerner, Segonds, and Verdet, *Nicolas Copernic*, 1: 498.

40 Tiedemann Giese himself also composed an apologia intended to defend Copernicus's theory from theological opposition, entitled "Hiperaspisticon … ad Nicolaum Copernicum." The treatise is now lost. "Hiperaspisticon" means shieldbearer; see: Lerner, "Aux origines de la polémique anticopernicienne (II)," 425; Snobelen, "In the Language of Man," 700.

in which the latter had "aptly vindicated the motion of the earth from disagreement with the Holy Scriptures." Based on the same letter, it is believed that the *Epistola* was intended as an addition to *De revolutionibus*.[41] In the early 1970s, Reijer Hooykaas unexpectedly discovered the treatise in a compilation of early seventeenth-century pro-Copernican writings issued by the Antwerp/Rotterdam printer Johannes van Waesberge. He convincingly identified the unattributed and undated *Epistola de terrae motu* as the work of Rheticus.[42] Hooykaas's decisive argument for his attribution to Rheticus was the recurrent use of the phrase "my teacher," despite never mentioning Copernicus by name.[43] Rheticus was Copernicus's only known pupil, and he did in fact refer to Copernicus as his teacher in other places, such as in a letter to German theologian and close friend of Melanchthon, Wittenberg professor of natural philosophy Paul Eber (1511–1569), and in the *Narratio prima*.[44] In addition, the treatise referred to another account written by the same author in which he entered into astronomical detail, again without giving the specific title. This would be the *Narratio prima* in which Rheticus had announced his intentions to write a second account.[45] The opening sentence of the *Epistola*, quoted earlier, can be seen as functioning like a hinge between both treatises. Given Giese's reference to the treatise in July 1543, Kenneth Howell convincingly argues that Rheticus most likely wrote his letter between 1541 and 1543.[46] Rheticus's authorship has been questioned. In an unpublished Master's thesis, Shannon Higgins claims that the author could instead have been an early seventeenth-century Catholic, due, among other things, to striking similarities with pro-Copernican writings in a Galilean context.[47] Given the lack of conclusive evidence, however, her conclusion is rather a caution to further scholarship.[48] I cannot claim the final word on Rheticus's authorship myself, but deem it highly probable that Rheticus was indeed the author. There are Tiedeman Giese's reference in a letter to Rheticus; references by the anonymous author to his *praeceptor* and to another account that did what the *Narratio prima* did; the author's favorable disposition towards the heliocentric model of the universe;

---

41 Howell, *God's Two Books*, 58–59.

42 The Letter on the Motion of the Earth contains two references to the virtual scholarly community of the Republic of Letters, see: "Epistola," 53 (32), 64 (63).

43 For a complete overview of his argument, see: Hooykaas, *G.J. Rheticus' Treatise*, 17–19.

44 Rosen and Burmeister, "Rheticus' earliest extant letter," *Isis* 61 (1970): 384–386.

45 Rheticus, *Narratio prima*, E4^v, F4^v.

46 Howell, *God's Two Books*, 59.

47 Higgins, "G.J. Rheticus."

48 For a status quaestionis of the authorship debate, see: "Die Abhandlung De Terrae motu von Georg Joachim Rheticus," 70.

the semantical logic of a twofold account articulated in the first sentence of the *Epistola* that tied both writings together; and the use of Scripture as a touchstone which is reminiscent of Melanchthon's emphasis on the compatibility of natural philosophy with Scripture.[49] According to Geoffrey Blumenthal, the letter wasn't published together with *De revolutionibus* due to the highly tense religious-political situation in the Holy Roman Empire. Rheticus and Copernicus were informed at an early stage about the prospective failure of the Colloquy of Regensburg in 1541. German theologians and the papal legate failed to compromise and the articles on which they did agree were not supported by Luther nor the pope. The failure of Regensburg therefore made it highly likely that Emperor Charles v would take action against the Protestant territories like Nuremberg. A tightening of the Imperial Edicts on Censorship was bound to take place, making Protestant printers and cities more careful.[50] Let's not forget that on 21 July 1542, Pope Paul III (1468–1549) constituted the Roman Inquisition with the bull *Licet ab initio*, and on 12 June 1543 banned the printing and sale of all books unless authorized by the Catholic church.[51] In this atmosphere it was better not to publish Rheticus's *Epistola de terrae motu*, especially because it argued in favor of having a non-canonized philosophical opinion and how this position could be motivated from Scripture. The reference to the Catholic faith in the *Epistola* can be explained by a number of reasons, one of which involved not raising suspicion in times of censorship.

3.1 *Not a Course of Physics*[52]

In general, the sixteenth century witnessed an increase in the bottom-up importance of the Bible in scientific debates about nature conducted by non-theologians, and Copernican astronomy was no exception to this.[53] Rheticus's *Epistola* too discussed the status of the Bible as a cognitive category in cosmological debate. His main intention was to show that "he who assumes its [the earth's] mobility in order to bring about a reliable calculation of times and motions is not acting against Holy Scripture."[54] The Bible, according to the mathematician, is the unique source of infallible truth on salvation, but

49 The recent reference work by Michel-Pierre Lerner, Alain-Philippe Segonds and Jean-Pierre Verdet does not question the rightfulness of the attribution to Rheticus, see Lerner, Segonds, and Verdet, *Nicolas Copernic*, 1: 494–499.

50 Blumenthal, "Copernicus's Publication Strategy," 151–178.

51 Idem, "Diplomacy," 83.

52 "Epistola," 58 (47), 45 (8).

53 Donahue, "Solid Planetary Spheres," 248.

54 "Epistola," 60 (52): "Non igitur contra S. Scripturam facit, qui ad constituendam certam temporum et motuum rationem, ejus mobilitatem assumit."

God does not use the Bible to inform man about the particulars of the set-up of the universe. In the absence of revelation, the thus ignorant astronomer can only turn to two cognitive categories that support the use of Copernicus's hypotheses, these being mathematics and, still, some passages of Holy Scripture. Man is permitted to inquire into nature and into Scriptural passages referring to nature. Rheticus's main exegetical authority to substantiate this procedure is the Augustine of Hippo. On the threshold of the Council of Trent (1545–1563), the works of the bishop of Hippo were held in high regard amongst humanists, Catholics, and Protestant Reformers. To the latter, the Church Fathers exemplified a more authentic Christendom prior to the corrupting influence of the Scholastic theologians.[55]

In the domain of salvation, Rheticus argued, one had to acquiesce in God's Word without diverging from Scripture: "the true understanding of everything contained in Scripture is to be sought in Scripture, not elsewhere, either in reason, or in philosophy and the secular arts."[56] The *Sola Scriptura* principle is a first argument in favor of interpreting the *Epistola* as a Protestant treatise, the typical early Reformation emphasis on affirmation a second.[57] "Even if certain things seems repugnant to our senses and intellect," Rheticus wrote, "we should not doubt and suspend judgement like the Academics, but refrain our intellect and simply stand firm and acquiesce to the Word."[58] God's Word, wrote Rheticus, has the power of demonstration.[59] He explained the procedure of mere acquiescence by referring to the Fall and man's postlapsarian condition:

> After Adam's Fall, the knowledge of God had been progressively extinguished in our minds, and since, as Christ says, it is not the will of the Father that anyone should perish, God gave us His Word, that we might be sure concerning His will with regard to us, and might have always before our eyes what He wishes us to do, and how through the promised Seed He was to put away His wrath against us and receive us into His grace. This is what the Holy Spirit desired to reveal through the Word, and this is the

55 McGrath, *Intellectual Origins*, 44.

56 "Anonymous Treatise," 65; "Epistola," 43 (2): "[…] ex sacris literis vera notitia omnium eorum quae in iis continentur, sit petenda, et non aliunde aut ex ratione, aut ex philosophia et prophanis artibus accersenda."

57 Howell, *God's Two Books*, 64.

58 "Epistola," 43–44 (3): "Quare et si quaedam sensibus et intellectui nostri videbuntur repugnare, tamen non ideo dubitandum et academicorum more επεχειν, sed captivandum intellectum nostrum et simpliciter verbo standum esse atque acquiescendum."

59 Ibid., 43 (2).

> purpose of Holy Scripture. Therefore, all for whom the word of the Lord is a lamp to their feet, are bound to assent firmly to the articles of faith which the Holy Spirit, in order to attain the aforesaid end, proposes to us, and in no way let them be taken from us.[60]

The Bible was thus a lamp to the postlapsarian mind. On the authority of Augustine, Rheticus declared that such affirmation was not required in reference to Biblical passages about nature. In the former's unfinished *De Genesi ad Literam* (393–395), he had found that "the obscurities of nature which we perceive as the works of God, the omnipotent creator, are to be dealt with not by affirmation but by inquiry."[61] This is echoed in a statement made in the *Narratio prima* as well: God had left the universe to the discussion of man.[62]

> But as it is clearer than the day that God has left a good deal to our own efforts, so as to stimulate the arts and sciences necessary for life, and the things that pertain to education and the honest use of our mind, we should really follow in these things the thread of nature, by which first principles, reason and daily experience lead us. And, since God desires to be glorified in nature, there is no doubt that our study will be pleasing to Him. Therefore, He prompts the minds of great men to inquire into the nature which He created and He furthers and conducts their studies.[63]

---

60 "Anonymous Treatise," 66–67; "Epistola," 43 (3)–(4): "Cum post lapsum Adae notitia Dei in mentibus nostris indies extingueretur, ac ut inquit Christus, Non sit voluntas patris ut ullus pereat, dedit nobis DEUS verbum suum, ut de sua erga nos voluntate certi essemus, et perpetuò ob oculos haberemus, quid nos facere vellet, et quomodo per semen promissum iram suam super nos remissurus esset, nosque in gratiam recepturus. Hoc est illud quod Spiritus Sanctus per verbum revelare voluit, et hic est finis sacrae Scripturae. Proindè omnes, quibus quidem verbum Dominini est lucerna pedum suorum, tenemur ut articulis fidei, quos propter dictum finem nobis Spiritus Sanctus proponit, firmiter assentiamur, neque eos ulla ratione nobis patiamur eripi."

61 "Epistola," 43 (1): "De obscuris naturalium rerum, quae omnipotenti Deo artifice facta sentimus, non affirmando, sed quaerendo tractandum est [...]."

62 Rheticus, *Narratio prima*, F2^r.

63 "Anonymous Treatise," 71; "Epistola," 47 (12)–(13). "Cum autem Sole clarius sit, Deum nostrae industriae non pauca reliquisse, ut excitaret artes et disciplinas in vita necessarias, et quae ad cultum et ingenii honestam exercitationem pertinent, sequeremur vero in his natura filum, quo nos prima principia, ratio et quotidiana experientia ducunt. Et quia Deus vult in natura illustrari, non est dubium quinei hoc nostrum studium gratum sit. Ideoque et magnorum virorum animos ad inquisitionem naturae à se conditae exuscitat, et eorum studia provehit et gubernat."

Scripture was not "a course of physics": it deliberately failed to provide an exact description of nature.[64] The best proof was the absence of such knowledge among humankind:

> For who would deny that if God, the creator of nature, had desired to describe the nature of things for us by His Word, nothing in the whole world could have remained hidden to us, of which we now would not know exactly the nature, causes and powers? For who could better describe the eye than God, Who made it?[65]

The extent to which Holy Writ accommodated itself to the intellectual grasp of the people was limited moreover, and it did not always conform to the wisdom of this world. As a result, "because they have not been handed down by the Holy Spirit with that care as these things on which our salvation depends," these fragments received multiple interpretations.[66] The interpretations of these verses were subjective, meaning that they varied depending on someone's way of thinking, nature or education.[67] The exegetical principle of accommodation—*Scriptura humane loquitur*—was clearly used by Rheticus in order to motivate the validity of an alternative hypothetical set-up of the world.[68] Human ignorance and uncertainty thus prevailed, and therefore, the true believer has to acknowledge that these hypotheses are permissible products of human ignorance.

> Therefore, he [Augustine] instructs us in an excellent manner that we, who have been convinced by the truth, should say: Holy Scripture has not held this to be so, but rather human ignorance has perceived it.[69]

It is a fabric of ignorance that held the *Epistola* together and endorsed Copernicus's right to theorize.

---

64 Ibid., 58 (47), 45 (8).

65 "Anonymous Treatise," 71; "Epistola," 47 (12): "Quis neget si Deus naturae conditor verbo suo rerum nauram nobis epingere voluisset, nihil nos in toto mundo latere potuisse: cujus naturam, causas et δγναμεις non exactè teneremus? Quis enim melius descripsisset oculum, quam qui ipsum condidit Deus."

66 "Anonymous Treatise," 69; "Epistola," 45–46 (8)–(9): "[…] ut pote quae à Spiritu Sancto non ea cura sint tradita, ac in quibus salus nostra consistit […]."

67 "Epistola," 46 (10).

68 Snobelen, "In the Language of Man," 694.

69 "Epistola," 46 (10)–(11): "Itaque pulchre docet, nobis à veritate convictis dicendum esse: Hoc non habebat divina scriptura, sed senserat humana ignorantia."

In general, historiography of the Copernican astronomers' approach to Scripture has largely focused on their use of accommodation to strategically defend the permissibility of their stance. Stephen Snobelen has linked Rheticus's Biblical hermeneutics with other sixteenth- and seventeenth-century Copernican astronomers who applied the same line of exegetical reasoning. Accommodation, according to Snobelen, played a constructive role in the Scientific Revolution.[70] However, the early modern Copernicans' arguments that Scripture spoke in the language of man has often been interpreted as an insistence on a strict demarcation between the study of nature and religious affairs.[71] Likewise, the *Epistola* has been examined through this lens.[72] There is more, however, to the relevance of the accommodation argument in cosmological debate. Kenneth Howell has quite rightly pointed out that, for these early modern Copernican astronomers, the Bible was more than a deterrent which needed to be excluded from cosmological study, and thus more than a consequence of the socio-disciplinary power struggles between theologians, natural philosophers, and mathematicians. But what else, exactly, was it considered to be? Agreeing with Howell, Rheticus's use of the Bible was an expression of his religious identity.[73] The accommodation argument revolved around both knowledge and piety. The use of Scripture as a touchstone in cosmological debate was a proof of piety for it acknowledged the deficient postlapsarian condition of man and the Bible's status as infallible source of truth. Not reverting to Scripture was described as an intellectual vice, "a crime of sacrilege."[74]

Human doubt in inquiry however did have to acknowledge the limits of Catholic faith, as spelled out by Augustine.[75] First, man must affirm those things that are clear in Scripture and not cast doubt on passages that are self-evident or evident by comparison with other Scriptural passages, or explained by clear arguments. Secondly, the authority of Scripture must be maintained at all times when interpreting unclear passages.[76] Later in the *Epistola*, Rheticus indicated an exception to his hermeneutical principle, namely, when dealing with those things beyond man's experience. In the *Narratio prima*, he had already hinted at this by adducing a citation of Pliny the Elder (c. 23–79 AD):

70 Snobelen, "In the Language of Man," 730.
71 Westman, *Copernican Question*, 131.
72 Howell, *God's Two Books*, 67; Westman, *Copernican Question*, 131.
73 Howell, *God's Two Books*, 1.
74 "Epistola," 46 (10).
75 Ibid., 43 (1)–(2).
76 Ibid., 47 (11)–(12).

"To inquire what is beyond the universe or heaven, by which all things are overarched, is no concern of man, nor can the human mind form any conjecture concerning this question."[77] Only Holy Writ can give us knowledge here.[78] This stance is affirmed in the *Epistola* with respect to the supracelestial waters, hereby referring to Genesis 1:6 ("God said, Let there be a firmament in the midst of the waters, and let it divide the waters from the waters."):

> Holy Scripture testifies that there are waters above the firmament: Because these are placed beyond [the range of] the senses, we cannot be misled by the senses so that we would state otherwise, but we simply accept the Scripture.[79]

The universe that has been left to human discussion is the domain that lies within man's sensory reach. In the non-sensible domain, he had to accept the literal text of Scripture.

The Bible not being a textbook in physics was no license for random speculation. Even in an inquiry into nature, Holy Writ served as a lamp requiring from the philosopher that his theories would comply with the clear teaching of Scripture. Notice the parallel with Luther who considered the Bible useful in avoiding the pitfalls of human reason, and Melanchthon's exposition on the ancient philosophical schools and their respective compatibility with Holy Writ. Rheticus added this passage to underline that Copernicus's tenets should not meet with the same fate as those of Aristotle: Copernicus did not endorse conclusions that were incompatible with the Bible's clear teaching. Rheticus elaborated Luther's theological anthropology for the domain of the study of nature. Both the methods of affirmation and of inquiry imply the deficiency of the human intellect. The human mind was either expected to simply acquiesce or to formulate statements of ignorance before divine truth. This line of reasoning takes after Luther's argumentation that due to original sin the truths expressed in Scripture may not be rationally comprehended, nor should attempts be made to achieve rational comprehension by means of sec-

77 "Narratio prima," 144; Rheticus, *Narratio prima*, D2^r: "Mundi seu coeli, cuius circumflexu tegantur cuncta, extera indagare, nec interesse hominum, nec capere humanae coniecturam mentis."

78 Rheticus, *Narratio prima*, D2^r.

79 "Anonymous Treatise," 80; "Epistola," 52 (28): "Sacra Scriptura testatur supra firmamentum esse aquas: Haec quia sunt extra sensum posita, non possumus sensibus decipi, ut aliud statuamus, sed simpliciter Scripturae acquiescimus."

ular learning. The clear teaching of Scripture was to be accepted on grounds of faith. In the absence of affirmation, the faithful man proceeds by inquiry and ignorance.

What then is the purpose of Scriptural passages about nature if not to speak of these natural things like philosophers do? These passages served the salutary purpose of affirming divine creatorship and omnipotence, and the providential rule of the universe.[80] Rheticus credited these Biblical fragments with a threefold intention. The first is to remind the reader that nature is divinely created. Moses had written Genesis in order to impress on his people that the world was created by God from nothing:

> For since these people were by nature superstitious, and idolators dwelt all around them, He thought it necessary to provide in this way, that they should not depart from the Maker of the World and Creator of nature, and turn to worship the sun, the stars, or some other created thing instead of God, as Moses testifies.[81]

The second purpose is to underline that God wanted to become known through nature; it is therefore necessary to confirm that He made heaven and earth, and that there is no other god besides Him. The final purpose of these passages is to demonstrate His power and His fatherly care for humanity.[82] This, Rheticus wrote, is the clear teaching of Scripture.[83] Philosophical tenets incompatible with the doctrine of creation and everything that can be deduced from it (divine omnipotence and providence), such as Aristotle's argument about the eternal existence of the world, are to be rejected as impious and sacrilegious.[84]

3.2 *Limits of the Human Mind*

This subsection turns to the primary authority that endorses the use of Copernicus's hypotheses: human reason, or mathematics as Rheticus termed it. The genuine interplay between mathematics and natural philosophy will be elaborated when dealing with the *Narratio prima*. From the *Epistola*, Rheticus's views

80 "Epistola," 58 (47).

81 "Anonymous Treatise," 67; "Epistola," 44 (5): "Cum enim hic populus natura esset superstitiosus, et circum circa habitarent idololatrae, duxit ei hac ratione cavendum, ne derelicto fabricatore mundi, naturaeque conditore, se ad solem, stellas, aut aliquam aliam creaturam convertere et loco Dei coleret: Sicut testator Moses."

82 "Epistola," 44–45 (5)–(7).

83 Ibid., 45 (7).

84 Ibid.

can be deduced on the mind's postlapsarian potential for ontological truth, and its bondage to ignorance.

In the first section of the second part of the *Epistola*, Rheticus elucidated his view on the limitations of the human mind, and as such clarified the difference between valid and physically real hypotheses:

> We have sufficiently shown elsewhere that because the hypotheses regarding the motion of the earth accord so well with the phenomena, they can be exchanged with the defined as a good definition. For reason cannot determine other hypotheses closer to the truth, as the continuous passage of time has taught [us], and insofar as it is permitted to humans to have knowledge of these divine things; this is amply mathematically demonstrated by my venerable master in his work. And, therefore, the mobility of the earth can be considered an ascertained truth [...].[85]

Hypotheses are not truths: they are imitations of truth, their truth status is revisable and is informed by the limits of human reason faced with divinity. The compliance between effects and model prove that no astronomical model has been closer to truth, as is proven by the compliance with the effects (derivative truths).[86] Crucial in demonstrating Rheticus's take on man's intellectual fallibility and acknowledgment of the provisional character of the hypotheses is the following quote:

> That is why no-one should pretend that we assert the mobility of the earth to be true, unmindful of our weakness. What after all is more human than to stumble and be deceived?[87]

85 Ibid., 47–48 (14)–(15): "Satis alibi ostendimus, hypotheses de terrae motu adeo consentire τοις φαινομενοις, ut cum eis tanquam bona definitio cum definito converti possint. Neque enim ratio alias hypotheses veritati propinquiores, ut continua temporis successio edocuit, quantum quidem homini de divinis istis rebus scire licet, constituere potest, ut mathematicè à D. praeceptore in suo opere copiose ostenditur. Cum itaque pro comperta veritate terrae mobilitas habeatur [...]."

86 The passage quoted above is a principal argument in favor of Rheticus's authorship of the *Epistola*, and, in fact, is the decisive proof according to Hooykaas (see: Hooykaas, *G.J. Rheticus' Treatise*, 107). The first sentence is a nearly verbatim repetition of a line in the *Narratio prima*. Time is again referred to as the decisive element in judging the accuracy of a celestial hypotheses. The argument that it is closest to the truth resembles his quote from Aristotle's *Metaphysics* at the outset of the tenth chapter in the *Narratio prima* about the cause of derivative truths being most true. Finally, the venerated master seems very likely to be Copernicus, whom Rheticus in the *Narratio prima* also named as his preceptor.

87 "Epistola," 48 (15): "Quare nemo nos insimulet quasi nostrae imbecillitatis immemores ter-

This stumbling in particular takes place in the invisible realm of nature. A particularly interesting passage regarding Rheticus's outlook on the human postlapsarian intellectual condition is found in the conclusion of his exegetical discussion in favor of Copernicus's tenets in the *Epistola*. He repeated that exact knowledge about nature and other philosophical matters is attainable to the extent that God has permitted human reason:

> And since God wishes us to be better versed in matters divine than in corruptible nature, (Socrates' opinion should here be followed) we, men, know either too little or nothing about these divine matters—we see the world sustained in a most beautiful order, whence reason concludes that there is one first cause which sustains it all, namely God. But what God is, how He governs these things and what His will is towards us etc.: we penetrate none of this, except insofar as He willed us to know by His Word.[88]

Rheticus's words reveal that he believed God to be knowable *a posteriori* through the visible effects or phenomena of His creation, but on questions of the invisible aspects—what He is, how He governs these things, and what is His will towards humans—man acquired knowledge through Holy Writ alone. This distinction coincides with knowledge of God as creator and knowledge of God as redeemer. Subsequently, Rheticus delivered a similar argumentation for the spirits: "We perceive from effects that Spirits exist but what they are, we do not know. What kind they are, Scripture shows."[89] Again, human reason was seen to acquire knowledge only through effects and again, this to a limited extent, so that the existence of spirits must simply be affirmed. Knowledge about that which transcends the effects—what the spirits are—

rae mobilitatem ἀληθῶρ asseramus. Quid enim magis humanum quam labi et decipi." A similar statement is found in the *Narratio prima* linking the condition of being human with being mistaken sometimes and deceived even by the appearance of truth, this time, however, with respect to ancient theories of lunar eclipses; see: Rheticus, *Narratio prima*, C2^r.

88 "Anonymous Treatise," 80; "Epistola," 52 (29): "Et cum Deus nos magis velit circa divina versari, quam corruptibilem naturam, (Socratis sequendum erit judicium) nos homines de divinis istis rebus aut parum aut nihil scire, Mundum in pulcherrimo ordine conservari videmus, unde colligit ratio, unam esse primam causam ista omnia conservantem, nempe Deum. Quid autem sit Deus, quomodo haec administret, quae sit ejus erga nos voluntas etc.: nihil perspicimus, nisi quatenus ipse per suum verbum nos scire voluit."

89 "Epistola," 52 (30): "Sentimus Spiritus esse ex effectibus, sed quid sint, ignoramus. Quales sint, ostendit Scriptura."

was clearly believed to exceed the limits of human rationality and can be disclosed by God's Word only. Only aided reason may acquire such knowledge, if such knowledge is revealed at all. After the spirits, Rheticus turned to the celestial bodies.

> We see the sun, the stars and all the heavenly bodies rise, set and repeat their previous courses in a fixed plan [...] as for what the sun, etc., is, one says it is a lighted torch which is extinguished by night; another one says it is a fiery stone; some believe it is luminous water; Aristotle, to avoid saying nothing at all, calls the heavens and the stars the fifth Essence. To say in fact that it is no less certain that the sun does not rise than that it does rise, and that everything is equally uncertain, is a mark of insanity. As to whether the sun appears to us to rise because the earth stands still, or whether it has a movement of its own, we must say that reason and mathematics are for the former alternative, the senses, however, are for the latter. For this, thanks should be given to God, for He gives us the knowledge and the use of the effects of nature, which has been made, above all, for man's sake. So it is enough for us to have a method of calculation which corresponds exactly to the observed phenomena.[90]

The efficient causes of the motions of the sun and earth are aligned with the unknowable nature of the sun. It is their shared invisibility that renders them unknowable. This is not a radical skepticism. Whether one believed that the earth or the sun moved was a choice between reason and the senses. Based on an explicit reference to the Academics at the outset of the treatise, it seems highly likely that Rheticus was taking up again the theme of Academic doubt or suspension of judgment.[91] Rheticus thus affirmed that the truth of Copernicus's models was a metaphysical issue and hence out of reach of the limited scope of human reason, and therefore unsolvable; these hypotheses embod-

---

90 "Anonymous Treatise," 81; "Epistola," 52–53 (30)–(31): "Videmus solem, stellas, supera denique omnia certa ratione oriri, occidere, pristinosque cursus repetere [...] Quid sit sol etc. alius facem accensam, quae nocte extingueretur, alius lapidem ignitum, aliqui credunt esse aquam lucentem; Aristoteles ne omnino nihil diceret, appellat coelum et stellas quintam Essentiam. Dicere vero, non minus certum esse Solum non oriri quam oriri, omniaque pariter incerta esse, est hominum insanientium. An autem nobis oriatur sol terrae cessione, an ipsi insit motus, illud quidem concludit ratio, et mathematica; hoc vero sensus. Pro hoc agendae sunt gratiae Deo, quod largitur nobis notitiam et usum effectuum naturae, a se propter hominem ut quam maxime conditae. Sic satis est, nos habere calculum τοις φαινομενοις exacte respondentem."

91 "Epistola," 43–44 (3).

ied most-true causes insofar as they retained this compliance with the celestial effects, i.e., as they could continue to be affirmed on the level of derivative truths. That Rheticus presumed the unknowability of the metaphysical is substantiated by another of his remarks, which reads as follows:

> Indeed, our knowledge of the causes of natural things is so insignificant that even the natural philosophers have to admit it in the most trivial matters, as when they say that this herb has this power not due to such a proportion of heat and humidity but due to the specific nature inserted in it by God.[92]

Man cannot fathom divine creatorship entirely. The metaphysical, that which cannot be observed, *in casu* the metaphysical themes of substance and causality, is unknowable. Unless informed by revelation, humanity—astronomer or philosopher alike—remains ultimately ignorant about the invisible aspects of nature, whether this concerns the efficient causes of planetary motions or the essence of herbs. From this account it follows that ignorance was the rational incapability of postlapsarian man to grasp divine omnipotence beyond its visible expression. What is more important than truth is the applicability of the hypotheses. Near the end of the treatise, Rheticus used the well-known verse in Genesis 1:14 to explain the purpose God had in mind for the specific inquiry into heaven, notably a certain calculation of time.[93] The first chapter has shown that this purpose was not secular: the astronomers at Wittenberg University portrayed themselves as timekeepers of sacred time. Rheticus's emphasis on utility implies that the ability to have an adequate method of calculation was the key criterion for the adequacy of Copernicus's astronomical models. This was not an instrumentalist stance since he did say that these causes were most true: as such, it was an admission of the partial knowability of nature.

This ignorance of the invisible causes of the celestial phenomena is portrayed by Copernicus's pupil as a theological anthropological condition. Exhaustive causal knowledge was to arise with the second advent of Christ, when man had obtained divine eyes.

92 Ibid., 53 (31): "Verum enimvero quam sit exigua causarum rerum naturalium notitia, etiam physici in lenissimis quandoque rebis fateri coguntur, ut cum dicunt, haec herba non habet hanc vim propter calidi et humidi tale temperamentum, sed propter specificam naturam a Deo ei inditam etc."

93 Ibid., 60 (52).

> When we cast off this corruptible nature and the Kingdom of Christ dawns, we will see face to face, as Scripture testifies, and we will perceive all nature before us, and with divine not human eyes.[94]

The use of the words "divine eyes" re-emphasizes the connection between human knowledge and experience. Redeemed man will see what was previously not visible, he will have direct access to the meta-visible. Rheticus located this post-redemptive state of exhaustive cognition after the second advent of Christ on earth, meaning after the general Resurrection, when the faithful were supposed to be living an eternal life. Rheticus's discourse about knowledge echoes the suggestion made with regards to Luther: knowledge of God and knowledge of nature were inextricably connected. The deficiency of the human mind caused by the Fall had the same impact on both domains. Rheticus claimed that the astronomer, like any other man, has to be pleased with that which he can know before he acquires exhaustive knowledge: "Therefore, let us be content with those things we can know, in the meantime praising God and thanking Him for all His benefits, Whose Name be blessed. Amen."[95] Rheticus's affirmation of temporal causal ignorance is portrayed as a humble affirmation of the human condition. The notion of immediate knowledge of the celestial world in the afterlife was a common one, often given to quite vivid, literal interpretations. In his 1549 oration about Regiomontanus, Erasmus Reinhold spoke about the deceased astronomer's posthumous condition in the following terms:

> [...] let us consider that he now sees the idea of the entire world in God Himself, and the causes of the way in which everything is made, that he beholds in person the heavenly bodies, and that he converses most pleasantly with the first fathers about the invention of the arts and their observations, as with fellow philosophers, and worships God the creator.[96]

---

94 Ibid., 53 (31): "Ubi autem hanc corruptibilem naturam exuerimus et Christi regnum inchoabit, videbimus facie ad faciem, sicut Scriptura testatur, et totam rerum naturam, coram et divinis oculis, non humanis intuebimur."

95 "Anonymous Treatise," 82; "Epistola," 53 (32): "Proinde his quae scire possumus acquiescamus, laudantes Deum interim, et ei gratias agentes pro omnibus suis beneficiis, cujus nomen sit benedictum. Amen."

96 "On Johannes Regiomontanus," 246; Reinhold, "Oratio de Johanne Regiomontano," B1^v: "Quod cum fecerit Regiomontanus, nunc eum existememus et in ipso Deo cernere ideam totius mundi, et causas, cur hoc modo omnia condita sint, et coram aspicere coelestia corpora, et saepe cum primis patribus de inventione harum artium et ipsorum observationibus tanquam συμ φιλοφοις dulcissime colloqui, et Deum autorem celebrare."

Reinhold situated Regiomontanus's experiencing of this view at the moment of speaking, implying that, in his opinion, exhaustive knowledge was to be obtained after death.

The brief epilogue of the *Epistola de terrae motu* [(63)–(64)] brings us to the reason for the gap between a demonstrated (*a posteriori*) and ontological (*a priori*) truth, namely the unfathomable nature of divine wisdom.

> [...] let us pursue, (insofar as God allows us), the knowledge of how the Lord also wished these most perfect bodies of the world to be known by us. The philosophers say that some things are known to nature, but unknown to us. To this category let us indeed consign also disputes about hypotheses. For it appears that the Lord rightly said to Job: "Hast thou known the order of heaven, and dost thou set up its plan on the earth?"[97]

Rheticus referred to a distinction made by Aristotle. Absolute knowledge of the cause of an effect is knowable only by nature; accidental knowledge—its sensibly discernible particulars—is what is knowable to man who observes these particulars.[98] His epistemology built on the recognition that the world was created by God, and that the divine omnipotence displayed in this creation was only partly comprehensible to man. In this section, Rheticus turned the deficient postlapsarian condition of the intellect to the benefit of Copernicus's hypotheses: ignorance legitimates both Copernicus's novel claims and his departure from the traditional geocentric cosmology. Rheticus's epistemology exhibited conspicuous similarities to Luther's appraisal of man's theological anthropological condition after the Fall, to such an extent that it can be termed a Lutheran epistemology. His epistemological position is a translation into cosmological discourse of man's postlapsarian intellectual condition of metaphysical ignorance. To resume the terminology used by Luther in his commentary on Genesis, Copernicus's hypotheses could rise to the level of common truths, acquired by experience, but could not attain a universal character. Rheticus propounded the possibility of knowledge through experience but confessed ignorance regarding true knowledge of that which exceeds experience. God was knowable in His ordained power, materialized in visible

97 "Anonymous Treatise," 101; "Epistola," 64 (64): "[...] assequamur (quantum nobis à Deo conceditur) notitiam quomodo haec pulcherrima mundi corpora nobis Dominus insigna quoque esse voluerit. Philosophi dicunt, quaedam naturae esse nota, nobis autem incognita. Ad hoc genus referamus sane et de hypothesibus disputationes. Recte enim apparet Dominum dixisse ad Iob: 'Nunquid nosti ordinem coeli, et pones rationem ejus in terra.'"

98 Jardine, "Epistemology of the Sciences," 687.

creation as displayed before the eyes of man, but not in His absolute power. Hence, God's works were but partially knowable, limited by experience. In the absence of revealed truth about cosmology, human beings—astronomer and philosopher alike—were temporarily ignorant of the invisible efficient causes of the planetary motions. According to this somewhat unorthodox interpretation, the axiom of uniform and circular planetary motions is the final cause of the planetary motions, the matter of the phenomena the material cause, and the outward discernible particulars the formal cause, implying the size, shape, position and motion (sequence of positions) of the celestial bodies. The hypotheses are the invisible efficient causes that give an explanation as to why the planetary motions are seen to unfold in a particular way, and reconcile the observed motions with the axiom. Whereas research into the formal cause of a planet yielded information about its trajectory in heaven, the efficient cause explains why a planet is observed to behave in a particular way. The efficient causes are the missing link between the final and the formal cause, between the axiom and the observed behavior of the celestial phenomena. Rheticus's temporal causal ignorance implies that astronomy could not evolve into an *a priori* science: the astronomer was capable of *quia* demonstrations only during his life on earth. He experienced a theological anthropological condition of temporal underdetermination as long as he himself had not seen nature clearly with divine eyes. In short, Rheticus did not believe man to have obtained a *propter quid* demonstration of the celestial phenomena.

There is no doubt that Rheticus's moderate skeptical epistemology was rooted in earlier intellectual traditions. Besides nominalism, which was an integral aspect of Luther's epistemology,[99] Augustinian epistemology was probably significant for Rheticus in this regard. The writings about learned ignorance of the cardinal and mathematician Nicolas of Cusa (1401–1464), moreover, could equally have had a strong influence. Rheticus very briefly mentioned his hypothesis that the earth is also luminous and one of the stars, and referred explicitly to the latter's *De docta ignorantia* (1440).[100] The brevity of the reference, however, shows only that Rheticus knew of Cusa and the doctrine of learned ignorance.

---

99 Hockenbery Dragseth, "Introduction," 4.

100 "Epistola," 60 (51). Cusa's theory of learned ignorance is in fact an agnostic doctrine of ignorance. Wisdom means awareness of the limits of one's knowledge; learning, therefore, is learning what one does not know. God is a convergence of opposites and therefore, one cannot have true knowledge of God; see: Kenny, *Medieval Philosophy*, 105, 311, 312.

3.3 *A Biblical Hermeneutics of Obscurity*

We turn now to the second authority that Rheticus believed could argue for the permissibility of Copernicus's hypotheses, namely to passages of Holy Scripture.[101] The second and major part of the *Epistola* [(17)–(63)] renders an exegetical discussion in favor of the permissibility of Copernicus's hypotheses, as initiated in the introduction. Contrary to what has often been assumed, Rheticus did not set out on this exegetical quest with the intention of excluding Scripture from the study of nature nor in order to find references there to the physical reality of the heliocentric world system. He intended to show that the Bible did not teach unequivocally the geocentric set-up of the universe. Rheticus did not claim to have found proofs of the physical truth of the heliocentric model in the Bible.[102] His exegetical strategy regarding Scriptural passages about nature was held together by the same thread as his inquiry into nature itself: ignorance was both the motive that vindicated the inquiry as well as its final conclusion. In reference to the Biblical passages about nature, he developed what I term "a hermeneutics of obscurity." Both the subject pursuing this exegetical inquiry—the postlapsarian mind—and its object—the causes of the phenomena—were obscure, just like the Scriptural passages about nature. Rheticus's hermeneutical endeavors echo Luther's warning that, when reading the Bible, exegetes had to acknowledge their ignorance before divine truth. His was not a Biblical quest for knowledge but for piety. In the introduction to the second part of the treatise, encompassing the actual exegetical discussion, Rheticus described the purpose of his hermeneutics of obscurity in terms of piety.

> And because the mobility of the earth can be considered an ascertained truth, it is not to be feared that the mark of impiety has been attributed to us by more righteous and erudite judges; let us inquire nevertheless whether certain interpretations of the more obscure passages about nature in the Holy Scriptures can be advanced.[103]

Rheticus's exegetical discussion aims to prove that an adherent to Copernicus's theories who used the theories for the purpose of calculation was not acting impiously (and thus did not disturb his relationship with God) because

---

101 Ibid., 59 (49).

102 Hooykaas, *G.J. Rheticus' Treatise*, 107.

103 "Epistola," 48 (15): "Cum itaque pro comperta veritate terrae mobilitas habeatur; non est timendum, ut ab aequioribus et eruditioribus judicibus nobis nota impietatis ascribatur, tametsi inquiramus, num inde etiam interpretationes quaedam locorum obscuriorum in sacris literis de rerum natura sumi possint."

Scripture was unclear on the subject.[104] Ignorance, provoked by the absence of clear teaching of Scripture, was a full-fledged component of Rheticus's notion of intellectual piety. His comparison of Scriptural passages speaking of the sun and the earth showed that God's Word was veiled in obscurity on the subject of the constitution of the universe. Holy Scripture arose as a touchstone of piety for prior conclusions made by following the thread of nature as he deemed fit in the study of heaven, thus *after* having established the demonstrated truth of the motion of the earth.

The first subdivision [(17)–(31)] deals with the Scriptural passages which could be adduced in favor of the author's position, focusing on the immobility of heaven [(17)–(18)] and the mobility of the earth [(18)–(31)]. Rheticus never argued that his interpretations were final. Instead, he concluded his exegetical research rather tentatively:

> These are the passages of Scripture, by means of which we could say that, if the earth is moved, something of this matter, albeit obscurely, is contained in the Bible.[105]

His specific articulation may indeed have been chosen on strategic considerations, realizing that he was, as Hooykaas suggested, venturing on to dangerous grounds. Hooykaas took Rheticus's remark as a blatant display of inconsistency, arguing first that the Bible was not a physics textbook and subsequently turning to Scripture to look for Biblical passages in favor of his teacher's model of the universe.[106] According to Kenneth Howell, however, Rheticus's Scriptural endeavors in this part did tie in with the Protestant notion of the perspicuity of Scripture. Rheticus contended that Scripture was clear in matters of salvation and that other truths might lie hidden beneath Holy Writ's redemptive message. As a result, the quote did not harm Rheticus's position on the non-philosophical status of Holy Writ.[107] Even more, this seeming inconsistency is resolved when integrating Rheticus's hermeneutics within his discourse of ignorance in favor of Copernicus's heliocentric model. It should be borne in mind that Rheticus concluded this section by expounding upon the pre-redemptive human condition of ignorance about the invisible components of nature. In my view, this exposition stressed the non-decisive character of his

---

104 Ibid., 60 (52).

105 Ibid., 52 (27): "Hi sunt fere loci Scripturae, quibus si terra movetur, aliquid hac de re, sed obscure in bibliis possemus dicere, contineri."

106 Hooykaas, *G.J. Rheticus' Treatise*, 107.

107 Howell, *God's Two Books*, 65.

Scriptural discussion in favor of Copernicus's tenets. The truths underlying the superficial contents of the text were not expected to become readable.

The language used in the above citation is exemplary for the cautious attitude exhibited in the preceding hermeneutical discussion. Another example of such cautious phraseology employed by Rheticus is his figurative reading of Biblical fragments drawn from Job and Isaiah in favor of the daily and annual motions of the earth. The verse from Isaiah is 40:22: "Lift up your eyes on high, and behold who hath created these things, that bringeth out their host by number: he calleth them all by names by the greatness of his might, for that he is strong in power; not one faileth."

> Perhaps, the text in Isaiah should be properly interpreted as referring to the daily motion, when the Lord says: "Lift up your eyes to heaven and look upon the earth beneath, that the heavens will vanish away like smoke." When we assume that the earth moves in a daily rotation around its poles, we see, when a revolution is half completed, a part of the sky which, for us, had previously been under the earth.[108]

The citation from Job is likewise dealt with in a hermeneutical fashion, this time applied to the annual revolution of the earth. The verse in question is Job 9:6: "(He) Who shaketh the earth out of her place, and the pillars thereof tremble." Rheticus argues that the citation from Job "may be understood in this way" or "perhaps" the fragment from Isaiah "should be interpreted in line with Copernicus's hypotheses." The treatise holds several other examples of such prudent usage.[109]

As a result of the principal "unreadibility" of Holy Writ in terms of knowledge about nature, Rheticus did not single out one preferred hermeneutical method of interpretation. His discussion displays a mix of literal and figurative readings of the Biblical verses in question. For instance, he gave literal interpretations of Biblical metaphors that could refer to the immobility of the heavens, such as "the tabernacle that is free from all disquiet" from the book of David.[110] Another example refers to a figurative reading, in support of the mobility of the earth.

---

108 "Anonymous Treatise," 76–77; "Epistola," 50 (21)–(22): "De motu vero diurno terrae, locus Esaiae intelligendus fortasse erit, cum dicit Dominus, Levate in coelum oculos vestros et videte sub terra deorsum, quia coeli quasi fumus liquiescent. Cum ponimus terram circa suos polos moveri motu diurno, facta dimidia revolutione, videmus partem coeli, quae nobis antea erat sub terra deorsum."

109 Ibid., 48 (17)–53 (31).

110 "Anonymous Treatise," 73; "Epistola," 48 (17): "E tabernacula quod ab omni inquietudine [...]."

> What in heaven is more unfixed and mobile than the Moon? If, however "to found" signified "to make immobile," David would be saying that it [the moon], along with the rest of the stars, was immobile, when he says:— "When I see Thy heavens, the work of Thy fingers, the moon and the stars which Thou hast founded." In a similar way, God did not render the earth immobile, either by fixing or establishing it, for Scripture attributes the same to heaven, as when David says: "By the Word of the Lord were the heavens established […]."[111]

In reference to the verse from David (Ps.8:3; "When I consider thy heavens, the work of thy fingers, the moon and the stars, which thou hast ordained"), Rheticus negated the figurative interpretation of the verb "to found" as "to make immobile." He clearly pursued an individual reading of Scripture. In the obscurity of Holy Writ, and of the human mind, for that matter, the individual conscience of the interpreter is presented as the deciding authority concerning Rheticus's alternative interpretation.

> Notwithstanding, I affirm nothing, that is to be followed what seems to be true to somebody.[112]

It was up to the individual inquirer into nature to decide what seemed true to him. He subsequently explained his own choice by arguing that he preferred mathematics over the senses when it came to deciding whether or not the sun moved. Another reference to the individual conscience of the interpreter as decisive for the choice of interpretations is given near the end of the treatise, after a final exposition on a Scriptural reference to the "trembling" of the earth.

> As far as I am concerned, anyone may interpret this passage figuratively, or in whatever way. What to the learned and pious has seemed the most suitable assumption in all this debate, we, too, shall follow.[113]

111 "Anonymous Treatise," 94; "Epistola," 59 (49): "Quid magis in coelo vagum seu mobile, quam luna, tamen si fundare esset immobilem reddere, et David eam cum reliquis stellis dixisset immobilem, cum inquit: Quoniam videbo coelos tuos, opera digitorum, lunam et stellas, quae tu fundasti. Simili modo Deus, neque firmando neque stabiliendo terram immobilem effecit, idem enim Scriptura coelo tribuit, ut David; Verbo Domini coeli firmati sunt […]."

112 "Epistola," 52 (28): "Nihil tamen affirmo, quod cuilibet visum fuerit, sequatur."

113 "Anonymous Treatise," 101; "Epistola," 64 (63): "Verum per me quispiam hunc locum poterit figurate dictum intelligere aut quomodocunque. Quod doctis ac piis in tota hac disputatione visum fuerit convenientissimum ad assumendum, nos quoque sequemur."

The combination of Rheticus leaving the decision up to the individual reader together with such cautious phraseology shows that he did not intend to set forth a decisive interpretation but proceeded by uncertainty instead. The obscurity of Holy Writ left room for other non-standard interpretations to emerge: Rheticus put forward one alternative interpretation based on previous inquiry into nature. The strong appeal to the individual conscience of the reader, in particular, is characteristic of a Protestant Biblical hermeneutics that, by removing the decisive impact of tradition and ecclesiastical authority, facilitated and vindicated an individual interpretation of Biblical verses. Simultaneously, Rheticus mentioned that he is prepared to follow what the learned and the pious deemed most suitable. In another instance, the learned and the pious are described as the church.[114] Instances such as these explain the confusion over the confessional self-identification of the author of the *Epistola*.

The second part of Rheticus's Biblical hermeneutics went deeper into the Scriptural passages that could be adduced by opponents aimed at countering his position. These fragments center around the mobility of heaven [(33)], the immobility of the earth [(34)–(52)] and the mobility of the sun [(52)–(63)]. The mere fact that the *Epistola* offers an alternative interpretation of Biblical verses whose meaning in ecclesiastical tradition was considered fixed demonstrate that its author is much more likely to have been active within a Protestant confessional environment. Rheticus was essentially challenging the canon here. His line of reasoning again ensues from the original point of departure, namely that the Bible was not a course of physics. The accommodation principle, introduced by him in the prologue, was central in arguing for the permissibility of his point. This exegetical principle entails that Holy Scripture was written in the language of man, and thus that its usage was adapted to the intellectual and spiritual level of humankind.[115]

> Many passages of Scripture could be collected by way of showing that Scripture often accommodates itself to popular understanding, and does not seek exactness in the manner of the Philosophers.[116]

114 "Epistola," 48 (16).

115 Benin, *Footprints of God*, xiv.

116 "Anonymous Treatise," 87; "Epistola," 55 (39): "Multi possent colligi scripturae loci, quibus ostenderetur, Scripturam se saepe accommodare ad vulgi captum, neque το ακριβες philosophorum more quaerere."

The necessity of accommodation is caused by the unbridgeable gap between the understanding of God and His people. Rheticus justified the use of this principle on the authority of Augustine and of Nicholas of Lyra (c. 1270–1349), a late-medieval Franciscan who had written one of the most widely read and copied commentaries on the Old and New Testaments, known as the *Postillae*.[117] Lyra's commentary is seen an important step in the preparation of the Reformation. The first and major part of the *Postillae* was a commentary on the Bible in its literal sense. The connection between the medieval exegete and the Reformation has often been articulated in the following terms: "If Lyra had not played the lyre, Luther would not have danced."[118] Lyra's name appears more than once throughout Rheticus's treatise. One particular example is the following.

> So, on the authority of Nicholas of Lyra, it was because of the uncultivated state of the people that, in the beginning of Genesis, no mention is made of the air, much less of the element fire, as being beyond the perception of the uneducated.[119]

In Rheticus's opinion, the poor level of Israelite erudition is the reason why the Bible did not give information about the planets besides the sun and the moon, and the reasons why other things have been left out as well. Likewise, the vacuum referred to in Job 26:7 ("Who stretcheth out the north over the empty place") is not to be understood in an Aristotelian sense. Nor is a reference in the New Testament to growing dead seed to be interpreted in a natural philosophical manner. Natural philosophers would certainly deny that dead seed grows into a plant and bears fruit.[120]

Recognizing that Scriptural language was accommodated to the intellectual level of mankind implies that Biblical references literally signifying the motion of the sun are to be seen as common figures of speech, as statements of obvious experience and not of learned observations. Phenomenalism was an integral tool of Rheticus's accommodationist hermeneutics.[121] Biblical usage regarding nature was tailored to the level of the unlearned or naïve observer.

117 Smith, "Nicholas of Lyra," 49.

118 Ibid., 50–51: "Si Lyra non lyrasset, Lutherus non saltasset."

119 "Anonymous Treatise," 87; "Epistola," 56 (39): "Sic propter ruditatem populi, ut Nicolaus de Lyra author est, in principio Genesis nulla facta est mentio de aere, multo minus de elemento ignis, quia non incurreret rudioribus in oculos."

120 "Epistola," 56 (40).

121 Snobelen, "In the Language of Man," 702.

This argumentation is of particular interest to a Copernican astronomer whose physical claims imply that the immediate perception of rotating heavens and an immobile central earth is an optical illusion. Rheticus argued that the perception of motion was provoked either by a motion of the perceived object or by a motion of the viewer of that object. Common speech, however, did not distinguish these types of motion. Among other examples, he referred to the ship that sails from harbor, causing people to say that the land and the towns recede from them. Even if the rational conclusions prove to be at variance with sensory observations, man still adopted the accepted ways of speaking. Rheticus argued that in everyday life, man spoke of appearances and not of truth. For instance, the sun is said to rise and set "even though we hold these things true only in appearance, since our reason holds it to be immobile."[122] Rheticus insisted that the application of an accommodationist hermeneutics on the subject of the sun's immobility did not go against the usage set forth in Scripture. Copernicus's hypotheses neither reject the sun as the natural source of light and God's administrator, nor deny the obvious experience that the sun regulates the seasons.[123] The hypothesis of the immobility of the sun is not contrary to Scripture since Holy Writ does not assign any motion to the sun.[124] Scriptural references to solar motion are referring to the apparent movement of the sun. Rheticus indicated that his and his teacher's assumptions were not the assertions of an unlearned observer: they distinguished between appearance and reality because they were in search of the truth about things.[125]

As Westman points out, the application of the accommodation principle in the *Epistola* indeed allowed Rheticus to philosophize freely, but this does not imply that he argued for a complete separation of the domains of confession and the study of nature.[126] Instead, he propounded the recognition of accommodated language as beneficial for the popular credibility of Holy Scripture itself. The failure to recognize that Scripture was not a textbook in natural philosophy could discredit and damage the authority of the Bible.[127] Proceeding from an exclusive literalist interpretation of the Bible, and so using the Bible as a source of knowledge about nature, was unwarranted. A historical precedent

122 "Epistola," 62 (58): "[...] etiamsi teneamus haec de apparentia motu esse vera, cum ratio ponat eum immobilem [...]."

123 Ibid., 61 (56).

124 Ibid., 62 (59).

125 Ibid., 62 (58).

126 Westman, *Copernican Question*, 131.

127 "Epistola," 56 (40).

of unwarranted literalism was Lactantius, a fourth-century Christian author and otherwise a man of great learning and eloquence, Rheticus added, who had ridiculed others for believing, based on Scriptural passages, that the earth had a spherical shape.[128] The argument is very similar to Copernicus's reference to Lactantius in the dedication of *De revolutionibus* where he was characterized as exemplary of those men who shamelessly distorted the meaning of a Scriptural passage to suit their own purpose.[129] According to Rheticus, Augustine legitimized inquiry into Scriptural passages of nature so as not to deter "the learned and the wise of the world" who would otherwise have less faith in Scripture.[130]

The second part of the treatise extends the argument that Scripture could not have been written in accordance with "the wisdom the world."[131] The new territories opened up by the Portuguese voyages of discovery, for instance, had been proven to contradict descriptions of the earth mentioned in Scripture and in ancient geography. After the example of Ptolemy, the common geographers depicted water on globes or maps of the world where the Portuguese had first-hand experience of the existence of fertile territories. Likewise, the "Psalmist," meaning David, adapted himself to "popular opinion and understanding" and his words reflected the experience of his time.

> Just as the general run of cosmographers, on globes or maps depicting the whole of the earth, portray seas wherever no land has been explored, so e.g. Ptolemy believed that there were waters where the Portuguese have discovered in their maritime voyages that very fertile territories exist; hence the Psalmist, accommodating himself to popular opinion and understanding, sang of the lands founded above the waters [...].[132]

The lack of knowledge about such matters shows that God did not intend Scripture to provide us with physical truths. Otherwise, Rheticus's argument goes, we would have known these things:

---

128 Ibid., 54 (35).

129 Copernicus, *De revolutionibus*, 4^{v}.

130 "Epistola," 46 (10).

131 Ibid., 45 (8).

132 "Anonymous Treatise," 86; "Epistola," 55 (38): "Quemadmodum vulgus cosmographorum, in globis, aut mappis, totum terrarum orbem referentibus, maria pingunt, ubi nullam terram exploratam habent, sicut Ptolemaeus credidit ibi aquas esse ubi Portugalensium navigationibus deprehensum est, esse fertilissimas regiones, ita Psalmista ad vulgi opinionem et captum se accommodans, terras super aquas esse fundatas cecinit."

> If the Holy Spirit had thought it necessary for salvation to tell that the earth was only the bed of the waters, who better than He could have been shown us by maps land known at that time, land explored in our time, and land to be discovered after some centuries, so as to convince us that lands everywhere lay beneath the waters? In fact, since the words of Scripture are concerned with sublime and totally different realities, they should not be alleged to censure the Philosophers.[133]

Likewise, he had argued in the introductory part of the treatise that we would have known all things about nature if God did not see inquiry as the preferred method in the study of both nature and Scriptural passages about nature.[134] Because Scripture spoke in human language and conformed to "the wisdom of the world," these Scriptural passages should neither be read literally nor be employed to censure natural philosophical opinions.

3.4 *Against Peripatetic Determinism*

Rheticus's argument against Biblical literalism hinged on the demand not to let human opinion determine the interpretation either of the celestial phenomena or of Holy Writ. Whereas in the first case, the phenomena cannot be saved properly, in the second case the meaning of the Biblical passage in question is distorted and the authority of Holy Writ itself is damaged. The underlying argument runs that the interpretation of Scriptural passages regarding nature ought not be tailored to the (then) level of erudition, and theories of planetary motion likewise should be allowed to evolve when required. In this part of the treatise, Rheticus affirmed once more that the Bible was not the book in which to look for knowledge about nature, but rather it teaches man about salvation: "the Holy Spirit has wished to write not a Natural Philosophy, but a rule of life, and to teach [us] how we are made children of God."[135] As a result, it was unwarranted to read Scripture as a book of physics. In the course of his discussion of the possible opposition to his position, Rheticus integrated a brief but serious criticism of the Peripatetics.

---

133 "Anonymous Treatise," 89; "Epistola," 56 (40)–(41): "Si Spiritus Sanctus judicasset necessarium ad salutem tradere, terram tantum alveum esse aquae, quis melius eo nobis in mappas retulisset, terram illis temporibus cognitam, nostro tempore investigatem, et post aliquot secula inveniendam, ut sic apertè nos convicisset, ubique infra aquas esse terras. Verum cum sese ad summa et longe alia respiciant verba scripturae, non erunt in Philosophorum censuram adducenda."

134 "Epistola," 47 (12).

135 Ibid., 58 (47): "[…] Spiritum Sanctum noluisse physicam, sed vitae regulam conscribere, et quomodo Filii Dei reddamur docere."

> Furthermore, there will not be lacking those who will bellow that it is monstrous to attribute movements to the earth, and who will take occasion to draw on and display their wisdom taken from the philosophers of nature. They are ridiculous, as if God's power could be measured by our force or our intellect. Are we to think that anything is impossible for God, who, by His Word, made the whole natural order out of nothing? Are we to tie God to the disputations of the Peripatetics about the heavy and the light, Him who is tied to no place, but who fills the whole world and is everywhere present and powerful, and who places the Antipodes firmly on the earth? And He who decreed that heaven should everywhere be above, is He unable to give the earth natural movements in accordance with its shape? Not in any way, in my opinion.[136]

In this respect, Scholastic natural philosophy was portrayed as a narrowing down of the rationally incomprehensible omnipotence of God to the intellectual capacity of Aristotle. This reproach builds on the same fundamentals as Rheticus's recommendation that the Bible not be read as a textbook in natural philosophy: one should maintain the authority of God's Word by not presuming the ability to rationally understand either Him or His Word. Seen in this light, the principle of accommodation serves its original purpose, namely as a recognition of the Jewish and Christian conviction on the one hand of the unfathomable nature of God's Works, and on the other of the limited intellectual abilities of man. Instead of being an exegetical tool of scientific progress wielded by the Copernicans, its use serves to acknowledge the limitations of the human mind (and thus the unreadability of Holy Writ in terms of knowledge about nature) and its tendency to conjecture impiously and yield absurdities before true doctrine. Although not named as such, the Peripatetics's impious failure to acknowledge God's infinite power devolved into a sort of intellectual vice, as the result of not recognizing one's postlapsarian condition. This type of determinism or preconceived judgment was unwarranted from a theological anthropological perspective: it was not in accordance with the world's

136 "Anonymous Treatise," 91; "Epistola," 57–58 (44)–(45): "Porro non deerunt, qui monstrosum esse vociferabuntur, terrae motus attribuere, et hic suam sapientiam ex physicis proferent et ostentabunt. Ridiculi, quasi potentia Dei ex nostris viribus ac nostro intellectu sit metienda. An putabimus Deo, qui ex nihilo totam rerum naturam verbo condidit, quicquam esse impossibile? An alligabimus Deum peripateticis disputationibus, de gravi et levi, ut qui nullo loco alligatus totum mundum complet, ubique praesens et potens est, quique firmat antipodem super terram. Atque ordinavit, ut coelum sit undique sursum, non possit terrae tribuere motus naturales ejus figurae convenientes ? Nequaquam ut opinor."

divine origin. Contrariwise, this reproach reflects upon Rheticus as doing precisely the reverse: he did not advance his teacher's theory, which had been developed from an *a posteriori* perspective, as either the one true interpretation of the celestial phenomena or the one true interpretation of Biblical verses referring to celestial bodies. He seemed to make his case that he, unlike the Peripatetics, did acknowledge the incomprehensibility of God in both His visible manifestations—namely, His works and His Word. The above quotation presents a mirror-image of Rheticus's preferred practice, which was not to reduce divine omnipotence to the mind of Aristotle but to turn to the celestial phenomena themselves. His criticism of the Peripatetics' impious determinism contrasts sharply with his exposition on what I have called temporary causal ignorance, ending the first part of his exegetical discussion. Taken together, these expositions demonstrate that the limited intelligibility of divine omnipotence is foundational to Rheticus's argument in favor of the permissibility of Copernicus's hypotheses. It is used as a vindication of the astronomer's right to theorize as well as of the exegete's right to bring forth an alternative interpretation. Ignorance, or the absence of a clear teaching in the Bible, took the shape of intellectual liberty in interpreting God's Word and works, that is, on the precondition of not overstepping the boundaries of Catholic faith, as Rheticus stated several times in the *Epistola*.

4 *Narratio prima*

Bearing in mind Rheticus's sketch of man's postlapsarian condition of temporal causal ignorance, his emphasis on the provisionary nature of astronomical hypotheses, and his reassurance of the pious nature of his proceedings, throw a different light on the epistemological limits of the disciplinary role-reversal taking place in the *Narratio prima*. Copernicus's hypotheses on the triple motion of the earth (daily, annual, and the motion of declination) and a heliocentric universe were the yield of a fruitful obscurity of the mind.

4.1 *The Thread of Nature*

The *Narratio prima* shows what happened when Copernicus followed "the thread of nature," led by first principles, reason, and everyday experience.[137] The treatise attempts to vindicate Copernicus's disciplinary incursion into cosmology from the perspective of a Renaissance astronomer who preferred

137 "Epistola," 47 (13).

mathematical certainty to the authoritative opinions which, for Rheticus and a host of early modern inquirers into nature who had followed him, had traditionally built upon. In line with his temporal causal ignorance displayed in the *Epistola*, it is clear nonetheless that he perceived astronomy as an *a posteriori* science. The astronomer is portrayed as being guided by the celestial phenomena. His opinion on the obscurity of the causes of the celestial effects—the domain of natural philosophy—was voiced in mythological terms in the *Narratio prima*. He characterized man's desire for knowledge of the causes of celestial phenomena as highly problematic, referring to a Cimmerian darkness.

> Finally, as Aristotle points out at length in another connection, men by nature desire to know. Hence, it is quite vexing that the causes of the phenomena are nowhere else [than in astronomy] so hidden and wrapped, as it were, in Cimmerian darkness, a feeling which Ptolemy shares with us.[138]

The Cimmerians, a mythical people from Homer's *Odyssey*, lived in a land of perpetual darkness at the end of the world, in "the deep waters of the river Oceanus where lie the land and city of the Cimmerians who live enshrouded in mist and darkness which the rays of the sun never pierce neither at his rising nor as he goes down again out of the heavens, but the poor wretches live in one long melancholy night."[139]

The astronomer has but the celestial effects. The *Narratio prima* has a distinct focus on the observational aspect of astronomy. Although this is not the typically modern outlook on Copernicus, which privileges his theoretical innovations, Rheticus described his teacher as an observational astronomer, making observations at the University of Bologna, where he was a pupil, an assistant and an observer of astronomy professor Domenico Maria Novara (1454–1504). And said that, in Frauenburg, Copernicus compared "all the observations of previous writers with his own."[140] Mary Boas Hall already remarked on the oddity of the fact that much of the early praise of Copernicus credited his observational efforts while, in fact, the astronomer carried out but few

138 "Narratio prima," 167; Rheticus, *Narratio prima*, F3^{r}: "Postremo cum homines quod Aristoteles alibi pluribus ostendit, natura sua scire appetant, nae molestum est satis, quod nusquam aeque causae των φαινομενων sint abstrusae, atque, ceu cimmeriis tenebris, involutae, quod ipse etiam Ptolemaeus nobiscum testare [...]."

139 Homer, *Odyssey*, 87.

140 Rheticus, *Narratio prima*, A2^{v}.

observations.[141] Rheticus's claim that Copernicus compared his own observations with previous ones shows that astronomical observation was also, at the time, considered to be a historiographical activity, researching a compilation of historical observations and adding the results of one's own stargazing activities to this store of information. The *Narratio prima* is filled with references to discrepancies in observational results of astronomers compared over time that Copernicus's theories of motion would supposedly have solved. The observers to whom Rheticus alludes range from the ancient Greek astronomers Hipparchus (*fl.c.*260 BCE) and Ptolemy, and the Arab medieval astronomer al-Battani (c. 858–929) through to Regiomontanus. Rheticus explained that these deviations were caused not by the instruments used, as was believed, but by the absence of "a definite and completely self-consistent law."[142] It was not the ancient observations that were defective but the hypotheses themselves. Rheticus credited his teacher for having exceeded the hypotheses developed by his predecessors by having harnessed existing experiences into a single framework. He managed "to arrange in a certain and consistent scheme or harmony the series and order of all motions and appearances, marshaled on the broad battlefield of astronomy by the observations of two thousand years, as by famous generals."[143] The particular accomplishment of Copernicus's restoration of astronomy, as his only known pupil repeatedly insisted, was that he had developed hypotheses which harmonized the astronomical observations of *all* past and present ages, providing a valid explanation for the past behaviors of celestial phenomena which were bound to be confirmed by future observations.[144] The record of historical observations, heaven itself, and mathematical reasoning had convinced Copernicus that Ptolemy had failed to establish a theory of planetary motion that was perpetually valid.[145] Although he had contributed greatly to the harmonizing of ancient observations recorded in the five hundred years predating him, time had disclosed that he had not succeeded in establishing a perpetual and consistent connection and harmony of celestial phenomena.[146] Rheticus described time as the true god and teacher of the celestial laws for "an imperceptible or unnoticed error at the founda-

141 Boas Hall, *Scientific Renaissance*, 92.

142 Rheticus, *Narratio prima*, B3ʳ.

143 "Narratio prima," 131; Rheticus, *Narratio prima*, C1ᵛ: "[…] quod seriem et ordinem omnium motuum et apparentiarum, quem observationes ii M. annorum tanquam praestantissimi duces in latissimo astronomiae campo explicant, in certam sibique mutuo consentientem rationem seu harmoniam colligere cogitur."

144 See, for instance: Rheticus, *Narratio prima*, A3ʳ, B1ᵛ, D1ᵛ.

145 Ibid., C1ʳ.

146 Ibid., C1ᵛ.

tion of astronomical hypotheses, principles, and tables is revealed or greatly increased by the passage of time."[147] Copernicus's hypotheses were advocated as being perpetually valid and, as such, surpassing the tables made by Hipparchus, Ptolemy, Theon of Alexandria (*fl.c.*364), al-Battani, and al-Zarqali (1028-c. 1087), and the Alphonsine Tables, which had served their purpose temporarily, for about two hundred years.[148] Now astronomical observation and theory concurred so well by means of Copernicus's hypotheses that he was said to have established a perpetual law: "The hypotheses of my teacher agree so well with the phenomena that they can be mutually interchanged, like a good definition and the thing defined," a clear intertextual reference to the *Epistola*.[149] Therefore, he continued, the motion of the earth may be considered an ascertained or demonstrated truth.[150] Hypotheses thus serve as the geometrical devices by means of which the observed celestial phenomena can be explained.[151]

The first principle that led Copernicus in his quest for an explanation of the celestial phenomena was the ancient axiom of uniform and circular planetary motions. The ideal materialization of the axiom was a system of homocentric spheres. Rheticus denounced the ancients for having neglected the axiom that all motions of heavenly bodies were either circular or composed of circular motions. One of his six reasons for abandoning the hypotheses of the ancient astronomers was precisely this neglect.

> Fourthly, only on this theory could all the circles in the universe be satisfactorily made to revolve uniformly and regularly about their own centers, and not about other centers—an essential property of circular motion.[152]

147 "Narratio prima," 132; Rheticus, *Narratio prima*, C1^{v}: "Siquidem insensibilis, vel etiam neglectus error, in principio constitutionis hypothesium, praeceptorum et tabularum astronomiae, procedente tempore sese aperit, aut etiam in immensum propagatur."

148 Rheticus, *Narratio prima*, B3^{r}.

149 "Narratio prima," 186; Rheticus, *Narratio prima*, H2^{r}: "[...] praeceptoris meis hypotheses τοις φαινομενοις consentire videbis, ut etiam inter se, tanquam bona definitio cum definitio converti possint." See "Epistola," 47 (14).

150 "Epistola," 48 (15).

151 Gingerich, "From Copernicus," 514.

152 "Narratio prima," 137; Rheticus, *Narratio prima*, C3^{v}: "Quarto hac unica ratione commode fieri posse D. preceptor videbat, ut quod maxim proprium circularis motus est, omnes revolutiones circulorum in mundo aequaliter et regulariter super suis centris, et non alienis moverentur."

Rheticus referred to the use of epicycles, deferents, and eccentrics in Ptolemaic astronomy that disturbed the picture of circular motions. Copernicus on the other hand had succeeded in reducing celestial motions to a pure combination of circles.[153] The former neglect of the axiom nearly caused the collapse of astronomy, in Rheticus's opinion. This axiom was seen by him as a cosmological principle, representing the "absolute system" of the universe:[154]

> Sixthly and lastly, my teacher was especially influenced by the realization that the chief cause of all the uncertainty in astronomy was that the masters of this science (no offense is intended to divine Ptolemy, the father of astronomy) fashioned their theories and devices for correcting the motion of the heavenly bodies with too little regard for the rule which reminds us that the order and motions of the heavenly spheres agree in an absolute system [...][155]

The planetary motions were by nature circular and regular. Saving the phenomena was interpreted here as saving the real uniform planetary motions.[156] The centrality of this axiom to the early reception of Copernicus's astronomy is clear from other sources as well. At the bottom of the title page of his copy of Copernicus's *De revolutionibus*, Erasmus Reinhold wrote: "The axiom of astronomy: celestial motion is circular and uniform or made up of circular and uniform parts."[157]

4.2 *Leaving the Tradition Argument*

Rheticus's mathematical preference for Copernicus's hypotheses stemmed from the fact that the allowed a reconciliation of all recorded celestial observations with the axiom of the uniform and circular planetary motions. His models

153 Gingerich, "From Copernicus," 515.

154 Lerner, "Origin," 411–412.

155 "Narratio prima," 138; Rheticus, *Narratio prima*, C3–[C4v]: "Sexto et postremo hoc maxime D. Doctorem praeceptorem meum movit, quod praecipuam omnis certitudinis in astronomia causam esse videbatur, quod huius doctrinae artifices (quod venia Divini Ptolemaei astronomiae parentis, dictum volo) suas theorias, et rationes motus corporum coelestium emendandi, parum severe ad illam regulam revocaverunt, quae ordinem et motus orbium coelestium absolutissimo systemate constare admonet."

156 Krafft, "Physikalische Realität," 266.

157 Gingerich, "From Copernicus," 515. Owen Gingerich retrieved Reinhold's copy of *De revolutionibus* in the Crawford Library of the Royal Observatory in Edinburgh.

provided the "best fit" between the celestial effects and their ultimate cause. As was typical of a Renaissance astronomer, Rheticus propagated the indispensable propaedeutic role of the mathematical disciplines for the study of heaven. Already in his inaugural lecture of 1536, the mathematics professor underlined that without geometry and arithmetic there would be no access to "that most excellent part of philosophy dealing with heavenly matters."[158] He affirmed the exceptional certainty of arithmetic, stating that "the art originates from the very nature of the human mind and produces the most certain demonstrations."[159] Rheticus's statements were very similar to those made by Regiomontanus in his inaugural lecture at Padua University in 1464. Regiomontanus had equally claimed the propaedeutic role of mathematics for the arts in general and natural philosophy in particular, arguing that someone who had neglected the quadrivium could by no means understand Aristotle's books on physics or meteorology.[160] Rheticus did not conclude, as Regiomontanus had or as Tycho Brahe would, by repudiating Scholastic natural philosophy, but he was in the process of extricating himself from the authority argument.

The generation of astronomers following Copernicus was now beginning to abandon a supposed canonical Aristotelian natural philosophy.[161] Copernicus himself had tried to anticipate the charge of novelty. In the dedication of *De revolutionibus* to Pope Paul III, he argued that the lack of certitude in traditional mathematical astronomy had prompted him to reread the books of all possible philosophers within reach. He found references to a possible motion of the earth in the writings of Cicero and Plutarch who respectively referred to Hicetas of Syracuse (c. 400–335 BCE) and to the Pythagorean philosophers Philolaus (*fl.c.*420 BCE), Heraclides Ponticus (*fl.c.*500 BCE), and Ecphantus the Pythagorean (c. 338-c. 315 BCE).[162] At the outset of his exposition on the triple motion of the earth, Rheticus likewise stated that his teacher adopted the opinions of Plato and the Pythagoreans.[163] In spite of his explicit reference to these ancient philosophers, Rheticus prioritized Copernicus's accomplishment to reconcile *all* historical records of observations of celestial phenomena with the "best fit" above attempts at compliance with

---

158 *In arithmeticen praefatio*, [A8^v].

159 Ibid., [A6^r]: "Ars oritur ex natura ipsius mentis humanae, et habet certissimas demonstrationes."

160 Swerdlow, "Science and Humanism," 148.

161 Jardine, *Birth of History and Philosophy of Science*, 189.

162 Copernicus, *De revolutionibus*, 4^r.

163 Rheticus, *Narratio prima*, D3^v–D4^r.

ancient authority. His vindication of the reasons why the Polish astronomer had chosen to develop new hypotheses in the first place shows an early abandonment of natural philosophy as a discipline that rightfully proceeded by authoritative opinions.

Although Rheticus's line of reasoning represents a disassociation from the Scholastic view, his words also tell us that he still functioned within a tradition- or canon-bound intellectual context.[164] Rheticus added that Plato and Ptolemy would have approved of Copernicus's ideas. He appealed to Plato in referring to the fact that geometry and the studies that accompany it dream about being, but the clear waking vision of it is impossible for them as long as they leave undisturbed their assumptions and cannot give any account of them. According to Rheticus, the great ancient philosopher would add that we must be deeply grateful to the immortal gods for the knowledge of such a theory of the phenomena.[165] Likewise, Ptolemy would approve of Copernicus's proceedings. When faced with the "royal road made inaccessible and impassable by the ruins of so many centuries," "he would have searched for another road across land and sea" to enable a sound science of celestial phenomena.[166] Rheticus vindicated his teacher's contravention of the opinions of Ptolemy and Aristotle from an *a posteriori* perspective of the study of heaven. Rheticus described astronomy and natural philosophy as inductive disciplines, proceeding as far as possible from effects and observations to principles.[167] Essentially, Rheticus argued that it was not opinions but experience and mathematics that were foundational for the study of heaven. Incongruity between the hypotheses and the actual observations of heaven had required Copernicus to opt for change, not some whim of the mind. And it is the same incongruity which rendered this change at least legitimate and non-foolish, as far as Rheticus was concerned:

---

164 Westman, "Astronomer's Role," 62. In the eighth chapter, Copernicus's pupil enumerates six principal reasons why the hypotheses of the ancients were to be abandoned. Besides three astronomical arguments, he refers to the axiom and the absolute system, two arguments that were discussed above. Another reason concerns God's skill as a clockmaker and seems to be a general argument rather than one specifically applicable to Copernicus's physical assumptions. This chapter has been subject to a rather elaborate discussion by Westman; see: Westman, *Copernican Question*, 121–123. We will therefore focus instead on some other examples of Rheticus's dealing with the argument of authority.

165 Rheticus, *Narratio prima*, D1^{v}.

166 Ibid., D1^{v}: "[…] regiam viam tot seculorum ruinis impeditam et inviam factam […] aliud insuper iter per terras mariaque inquisiturus esset […]."

167 Ibid., D1^{r}.

> Furthermore, concerning my learned teacher I should like you to hold the opinion and be fully convinced that for him there is nothing better or more important than walking in the footsteps of Ptolemy and following, as Ptolemy did, the ancients and those who were much earlier than himself. However, when he became aware that the phenomena, which control the astronomers, and mathematics compelled him to make certain assumptions even against his wishes, it was enough, he thought, if he aimed his arrows by the same method to the same target as Ptolemy. At this point we should recall the saying, "Free in mind must he be who desires to have understanding." But my teacher especially abhors what is alien to the mind of any honest man, particularly of philosophic nature: for he is far from thinking that he should rashly depart, in a lust for novelty, from the sound opinion of the ancient philosophers, except for good reasons and when the facts themselves coerce him.[168]

Rheticus's portrayal of Copernicus as experiencing something close to an inner conflict illustrates the controversial nature of the Polish astronomer's proceedings. The vindication rendered in the above citation is threefold. First, it was not a desire for novelty *per se*, but the celestial phenomena themselves which invited Copernicus to make this decision. Secondly, although Copernicus's hypotheses may deviate from Ptolemy's opinion, the Polish astronomer did respect ancient methodology. Earlier in the treatise, Rheticus wrote that Copernicus was convinced that "the rejected method of the sun's rule in the realm of nature must be revived, but in such a way that the received and accepted method retains its place."[169] Based on purpose and methodology, Rheticus still placed the Polish astronomer in a methodological tradition of

---

168 "Narratio prima," 186–187; Rheticus, *Narratio prima*, H2v: "Porro velim te, de doctissimo viro, D. doctore, meo praeceptore hoc statuare, tibique persuasissimum habere, apud eum nihil prius, nec antiquius esse quicquam, quam vestigiis Ptolemaei ut insistat: nec aliter, ac ipse Ptolemaeus fecit, veteres et se antiquiores multo secutus. Dum autem τα φαινομενα, quae astronomum regunt, et mathematica se cogere intelligent, quaedam praeter voluntatem etiam ut assumeret satis interim esse putavit, si eadem arte in eundum scopum cum Ptolemaeo tela sua dirigeret, etiamsi arcum et tela ex longe alio materiae genere, quam ille, assumeret: ac hoc loco illud arripiendum; δει δ'ελευθεριον ειναι τη γνωμη τον μελλοντα φιλοσοφειν. Caeterum, quod alienum est ab ingenio boni cuiuslibet, maxime vero a natura philosophica, ab eo qui maximum abhorret Dominus praeceptor meus: tantum abest, ut sibi a veterum recte philosophantium sententiis nisi magnis de causis, ac rebus ipsis est efflagitentibus, studio quodam novitatus, temere discendendum putarit."

169 "Narratio prima," 139. Rheticus, *Narratio prima*, [C4r]: "[…] damnatam rationem gubernationis in rerum natura solis, revocandam statuit: ita tamen ut receptae etiam et approbatae suus locus relinquatur."

astronomy that stretched back to Ptolemy and the ancient astronomers. He still embraced the immense authority of the ancient philosophers, albeit not on the level of their opinions but on the level of their methodology. Donahue points out that, throughout the sixteenth century in general, it was an obedience of the methods of the ancients rather than an adherence to their opinions that was expected.[170] The ancients would tacitly concur with Copernicus's new hypotheses, for they were driven by the sincere motives of a kindred spirit and themselves departed from the celestial effects when constructing hypotheses. Thirdly, Rheticus believed the astronomer to be entitled to some measure of philosophical freedom when reflecting upon the causes of the celestial phenomena.[171] The first public account of Copernicus's hypotheses was a legitimate exercise of the astronomer's right to dissent from the canonical account: "Free in mind must he be who desires to have under/standing" was the opening quote of the *Narratio prima*, inscribed on the title page, and was now recalled towards the end of the treatise.

This quote would become an "emblem of Copernicanism," reappearing also on the frontispiece of the fourth edition of the *Narratio prima* in 1596, published together in a single volume with the *Mysterium cosmographicum* by Johannes Kepler under the editorship of Michael Maestlin. It appeared also on the frontispiece of Kepler's *Dissertatio cum nuncio sidereo* in 1610.[172] While the quote reminded the reader of the astronomer's philosophical liberty when dealing with celestial phenomena, the problematic nature of this position was clear from the start. Achilles Pirmin Gasser (1505–1577), Rheticus's mentor who published a new edition of the *Narratio prima* in 1541 in Basel, wrote that this new cosmology was inevitably destined to be deemed contrary to the teachings of the schools and may even sound heretical, even if one day it would be accepted by all educated people.

Alcinous's quote was used as an explicit argument in favor of philosophical liberty, against a preconceived determinism based on the presumption of the intelligibility of the physical world. At the beginning of his ninth chapter, which deals with the arrangement of the universe, Rheticus reproached the natural philosophers for such opinion-based determinism:

> But the men who endeavor to pull all the stars together around in the ether in accordance with their own opinion, as though they had put

170 Donahue, "Solid Planetary Spheres," 247.

171 For more on the link between early modern Neoplatonism and the tradition of free inquiry, see Maclean, "'Sceptical Crisis' Reconsidered," 266.

172 Granada, "Travail of Philosophical Liberty," 63.

> chains upon them, merit pity rather than resentment, in your judgment as in that of other true mathematicians and all good men.[173]

Theories ought not to evolve into prisons or straitjackets of the mind. In the *Epistola*, Rheticus will be seen to nuance his position on the philosophical freedom of the astronomer by arguing for the human intellectual incapacity to grasp God's absolute power: his capacity was restricted to God's ordained power.

The notion of divine omnipotence will equally be seen to explain why Rheticus did not take the further step of equating Copernicus's model with full physical reality or ontological truth. Instead, he assumed the hypothesis of terrestrial mobility invoking the model to be "most true" or "very near the truth." Rheticus did not perceive astronomy as an *a priori* science capable of yielding ontological truth about the universe on grounds of fundamental human ignorance: he endorsed a probable, but not a positive claim about the constitution of the universe.

Several passages in the *Narratio* hint at a limited knowability of the world. One particular fragment, found in the tenth chapter, which discusses the planetary arrangement of the universe, recurs in the *Epistola de terrae motu* and is something of a pivot between the two treatises. The following citation concerns a line in Aristotle:

> Aristotle says: "That which causes derivative truths to be true is most true." Accordingly, my teacher decided that he must assume such hypotheses as would contain causes capable of confirming the truth of the observations of previous centuries, and such as would themselves cause, we may hope, all future astronomical predictions of the phenomena to be found true.[174]

When trying to understand Rheticus's position regarding the truth of Copernicus's planetary models, this quote has two major implications. The first is

---

173 "Narratio prima," 140; Rheticus, *Narratio prima*, [C4^{v}]: "Illud autem hominum genus, quod omnes simul stellas pro suo arbitratu, haud secus ac iniectis vinculis, in aethere circumducere conatur, commiseratione potius quam odio esse dignum, te iuxta cum aliis veris mathematicis, omnibusque viris bonis iudicare."

174 "Narratio prima," 142; Rheticus, *Narratio prima*, D1^{v}: "Aristoteles, inquit, verissimum est id, quod posterioribus, ut vera sint, causa est. Sic cum D. praeceptor meus, sibi tales hypotheses assumendas esse statueret, quae superiorum seculorum observationes ut versae esse confirmarentur, causas continerent, et quemadmodum sperandum, causae essent, ut in posterum omnes astronomicae των φαινομενων praedictiones verae deprehenderentur."

that the defining criterion for the veracity of hypothetical causes is apparently found on the level of derivative truths. Applied to Copernicus's planetary hypotheses, this means that his models were proclaimed to be most true because they were not at variance with the observed celestial phenomena. The second implication is that he considered the truth of such hypotheses to be a metaphysical issue which, in turn, explains why these hypotheses were most true instead of absolutely true. The specific citation is taken from Aristotle's *Metaphysics*.[175] The "first philosophy," as Aristotle labeled metaphysics, dealt with the first causes and first principles of things. The *Metaphysics* identifies the natural philosopher's search for causes with the metaphysical search for truth. Truth is the supreme cause of all things—representing the eternal and the absolute—and is situated at the end of the causal chain.[176] Aristotle's *Metaphysics* essentially seeks to explain how wisdom is to be acquired by our limited human capacities.[177] By initiating his discussion of Copernicus's arrangement of the universe with the above quote, Rheticus indicated that the planetary arrangement of the universe was a metaphysical issue. Reading his reference to the notion of "most true" in its Aristotelian meaning leaves sufficient room for some skepticism, pointing out that some degree of uncertainty persists due to the limited capacities of human understanding. At the outset of the chapter in *Metaphysics* from which Rheticus's quote was taken, Aristotle argued that human knowledge of truth is both difficult and straightforward at the same time, because "no one is able to attain the truth adequately, while, on the other hand, we do not collectively fail."[178] When discussing what truth meant to Rheticus, precisely this notion must be applied. Truth, or the knowledge of reality, is partly within reach through the effects which everyone can see, but truth in the sense of the infallible knowledge of the invisible causes is out of range.[179] Reality consists of visible and invisible components. Ontological truth implies actual existence, an unquestionable condition of being. Copernicus's hypotheses, by contrast, were most true. This interpretation is similar to Fernand Hallyn's reading of the support for Copernicus's heliocentric

---

175 Aristotle, *Metaphysics*, II, 1, 993b 26–27.

176 Lawson-Tancred, "Introduction," xliv.

177 This interpretation coincides with Hugh Lawson-Tancred's interpretation of the given citation from the *Metaphysics*, i.e., that Aristotle's use of "truth" is here closer to the concept of reality, since the philosopher argues that the truth of causes stands in relation to their being. If the causes have more being, they are more true, if less, they are less true; see: Lawson-Tancred, "Introduction," 3.

178 Ross, *Aristotle's Metaphysics*, 1: xxv.

179 Ibid., 1: xxv.

system articulated by the Louvain mathematician Gemma Frisius: observation and demonstration proved that the heliocentric model was the best available world system. This is not a claim to ontological truth but rather the best approximation of reality. According to Hallyn, Frisius's position was one of "prudent realism."[180]

5 The Historiographical Reintegration of a Once Lonesome Physical Realist

When using the Lutheran epistemological frame of reference sketched by Rheticus in the *Epistola* to analyze contemporary writings on the topic of Copernicus's hypotheses by other Lutheran authors such as Philipp Melanchthon and Andreas Osiander, their respective epistemological positions appear to align much more than previous historiographical polarizing might lead us to suspect. Rheticus's views on human ignorance and the Bible as a source of infallible truth recur to some extent in Osiander's anonymous preface to *De revolutionibus*, entitled *Ad lectorem de hypothetibus huius operis*, and in Melanchthon's textbook *Initia physicae doctrinae*.

5.1 *Andreas Osiander*

Rheticus met the Lutheran minister and Reformer Andreas Osiander in 1538 during the former's stay in Nuremberg as part of his academic tour.[181] The Nuremberg theologian with a strong mathematical interest had been involved in the coming-into-existence of the printing of *De revolutionibus*. In a letter sent to Rheticus on 13 March 1540, he had expressed his enthusiasm for the former's *Narratio prima*. Some three months later, in a letter dated 1 July 1540, Copernicus had asked Osiander for advice on how to present his astronomy relative to scriptural issues. He replied that theological opposition would be minimized if the astronomical theories were to be presented as mere hypotheses saving the phenomena and thus emphasizing the computational benefits.[182] After Rheticus's departure to the University of Leipzig, it was left to Osiander to supervise of the printing process of *De revolutionibus*.[183] With-

180 Hallyn, "Gemma Frisius," 83.
181 Barker, "Lutheran Contribution," 67.
182 Blumenthal, "Diplomacy," 81.
183 Burmeister, *Georg Joachim Rhetikus*, 1: 160; Wrightsman, "Andreas Osiander's Contribution," 230.

out informing either Copernicus or Rheticus, Osiander and Petreius decided to add an anonymous preface to the text. Blumenthal argues that Osiander did so because he probably didn't feel sufficiently confident to publish the text without provisos, given the tense religious-political climate due to the failure of the Colloquy of Regensburg, which seems highly likely. Protestant printers and commentators had become more careful.[184] Upon receiving the final print of *De revolutionibus*, Rheticus is said to have fallen into a rage over this preface.[185] His disagreement manifested itself in the copies of the first edition he presented to three friends in which he had crossed off Osiander's introduction on each page with an X in red crayon.[186]

Osiander's text has been revisited many times in the historiographical debate concerning the status of hypotheses in sixteenth-century astronomy, and is generally considered to be representative of the earliest technical or instrumental reception of Copernicus's astronomy in Lutheran circles.[187] It was a reading, according to Michel-Pierre Lerner, Alain-Philippe Segonds, and Jean-Pierre Verdet, that would eventually allow heliocentric astronomy the right to co-exist with geocentric astronomy at Reformed universities in the Holy Roman Empire, as long as the teaching did not concern the cosmological foundation.[188]

In the preface, Osiander denied that astronomical hypotheses could establish a physical claim about the constitution of the universe, vindicating Copernicus's hypotheses as the result of a conventional activity of hypothetical

184 Blumenthal, "Copernicus's Publication Strategy," 172.

185 A reference to this is found in a copy of *De revolutionibus* owned and annotated by the German astronomer Michael Maestlin who himself repeated what he had found among the books of another astronomer, Philipp Apian (1531–1589): "On account of this letter, Georg Joachim Rheticus, the Leipzig professor and disciple of Copernicus, became embroiled in a very bitter wrangle with the printer Johannes Petreius, who asserted that it had been turned over to him with the rest of the work. Rheticus, however, suspected that Osiander had prefaced it to the work; if he knew this for certain, he declared he would rough the fellow up so violently that in the future he would mind his own business and not dare to aggravate astronomers any more. Nevertheless, [Peter] Apian [father of Philip Apian] told me that Osiander had openly admitted to him that he had added this by himself"; see: Gingerich, *Annotated Census*, XVII. In the same section, Gingerich adduces other contemporary references that refer to Osiander as the author of the anonymous preface, amongst which is a reference by Kepler in his *Astronomia nova* (1609).

186 Gingerich, *Annotated Census*, 135–136. The three friends were Andreas Aurifaber, dean at Wittenberg, canon of Warmia Georg Donner, and Tiedemann Giese.

187 See, for example: Christianson, "Copernicus and the Lutherans," 6.

188 Lerner, Segonds, and Verdet, *Nicolas Copernic*, 1: 482.

model building.[189] From this perspective, predictive accuracy was the sole criterion for the adequacy of hypotheses:[190]

> For it is proper to the astronomer to draw up a history of celestial motions by means of diligent and skilled observation. Furthermore, since he cannot by any reasoning comprehend their true [causes], he must devise and construct their causes or hypotheses, by the assumption of which the same motions can be calculated properly from geometrical principles, both in the future and in the past [...]. And yet, it is not necessary that these hypotheses are true, or even probable, but one thing suffices—that they produce a calculation which agrees with the observations. [...] For it is sufficiently clear that this art [astronomy] is absolutely and profoundly ignorant of the causes of the apparent irregular movements.[191]

Michel-Pierre Lerner pertinently argues that Osiander wanted to highlight the "unreality" of these constructions by representing them as arbitrary human inventions.[192] There was no criterion to decide whether astronomical hypotheses were either true or false; they were selected on the sole basis of simplicity.[193] Unlike Rheticus, who had credited Copernicus's hypotheses as probable causes, Osiander proposed that they be seen as fictitious constructions, with no intentional crossings of disciplinary lines. By preventing a mathematical construction from evolving into a physical claim, Osiander conferred upon the astronomer the dubious right to proceed by absurdities that were not supposed to agree with reality.[194] He absolved the hypotheses from any claims to reality. Osiander's motivation was very likely to protect Copernicus's work

---

189 Gingerich, "From Copernicus," 514. Donahue regarded the preface as the reason why the physical claims in *De revolutionibus* were ignored during the phase of its earliest reception; see Donahue, "Astronomy," 568.

190 Jardine, *Birth of History and Philosophy of Science*, 212–219.

191 "Ad lectorem," 1v: "Est enim astronomi proprium, historiam motuum coelestium diligenti et artificiosa observatione colligere. Deinde causas earundem, seu hypotheses, cum veras assequi nulla ratione possit, qualescunque excogitare et confingere, quibus suppositus, iidem motus ex geometriae principiis, tam in futurum, quam in praeteritum recte possint calculari [...]. Neque enim necesse est, eas hypotheses esse veras, imo ne verisimiles quidem, sed sufficit hoc unum, si calculum observationibus congruentem exhibeant. [...] Satis enim patet apparentium inaequalium motuum causas, hanc artem penitus et simpliciter ignorare."

192 Lerner, "Aux origines de la polémique anticopernicienne (II)," 429–430.

193 Howell, *God's Two Books*, 45.

194 Hallyn, "L'absurdum àkroama de Copernic," 13.

from criticism.[195] Copernicus and Rheticus are said to have originally consulted him about his opinion on possible objections from the theological side.[196] In the only preserved excerpt of a letter addressed to Rheticus (20 April 1541), Osiander stated that Peripatetics and theologians alike would be easily placated if they heard that Copernicus's hypotheses were not certain and that different hypotheses could account for the same apparent motion.[197] The preface can thus be seen as a preemptive defense against anticipated objectives, arguing that the Polish astronomer had left the set-up of the liberal disciplines unharmed.[198] This was Osiander's strategy in favor of the astronomer's freedom to hypothesize.

Rheticus, however, legitimated the astronomer's freedom to philosophize independently by permitting him to make physical claims insofar as the theologically-anthropologically determined limitations of the human mind would allow, enabling as such a mathematical hypothesis to turn into a probable physical claim. What Rheticus and Osiander did agree upon in writing was the gap between the Bible as a source of infallible truth and the human mind, ever failing to attain such truth about the causes of the celestial phenomena. Osiander attested to this view by a comparison of the respective attitudes of the astronomer, the philosopher, and the Bible in relation to truth:

> When different hypotheses from one and the same motion present themselves (such as eccentricity and epicycles for the motion of the sun), the astronomer should rather adopt that [hypothesis] which is most easy to comprehend. The philosopher, perhaps, requires verisimilitude instead. But neither of them will understand or state anything certain unless that certain[ty] has been divinely revealed to him.[199]

Rheticus and Osiander did agree on the main epistemological result that man could achieve by himself, namely, the identification of most-true causes. Osiander expected that the study of nature would result, at most, in a *semblance*

---

195 Westman, "Copernicans and the Churches," 81; Likewise, according to Bruce Wrightsman, Osiander very likely chose to add the treatise anonymously in order to prevent a negative reaction from the Catholic Church due to his reputation as a Lutheran Reformer; see: Wrightsman, "Andreas Osiander's Contribution," 233.

196 Lerner, "Aux origines de la polémique anticopernicienne (II)," 427.

197 Andreas Osiander, letter to Rheticus, 15 April 1541, 3: 25.

198 Maclean, "'Sceptical Crisis' Reconsidered," 272.

199 "Ad lectorem," 1^v–2^r: "Cum autem unus et eiusdem motus, varie interdum hypotheses sese offerant (ut in motu solis, eccentricitas et epicyclium) astronomus eam potissimum arripiet, quae compraehensu fit quam facillima. philosophus fortasse, verisimilitudinem

of the real. True knowledge was obtained solely by revelation. Holy Scripture, in the words of Bruce Wrightsman, was the ultimate source of truth in his epistemological skepticism.[200] Barker and Goldstein take the above passage to imply that, for the earthbound astronomer, true causes are accessible by revelation only; in this light, they label Osiander's preface a restatement of the impossibility of moving beyond *quia* demonstrations in astronomy.[201] This line of reasoning indeed applies to Osiander's appraisal of the relation between human knowledge and revelation, but Osiander did not argue that Copernicus's hypotheses were posing as probable causes. In this light, we would have to agree with Michael Shank who opposed Barker and Goldstein's interpretation of Osiander's preface. In Shank's view, Osiander treated the hypotheses as fictitious models in line with the Ptolemaic tradition in mathematical astronomy.[202]

Like Osiander, Rheticus admitted the obscurity of the human mind unless lit by the clear teaching of Scripture. Osiander deemed that all claims to truth were to be measured against Holy Writ as the only source of truth.[203] This had been precisely Rheticus's intention in the *Epistola de terrae motu*: to measure Copernicus's claims against God's Word. This argumentation implies that hypotheses presuming to be causes were to stand the test of Holy Writ, the one source that could provide infallible truth in the invisible *a priori* realm of nature. By comparing Scriptural passages on the subject, Rheticus had established the absence of unambiguous clarity in this field. This obscurity equaled the obscurity of the human mind on the subject and allowed Copernicus's hypotheses to take the shape of most-true causes. By not granting Copernicus's models the doubtful status of being the actual causes of the celestial effects, Osiander made verisimilitude irrelevant—it was not a claim to truth—, and so avoided a confrontation between Copernicus's hypotheses and the Bible. Rheticus's *Epistola* had thus brought into effect what Osiander's preface attempted to prevent.

5.2 *Philipp Melanchthon*

Rheticus's views on human ignorance in the absence of revelation align with those of Melanchthon, just like their theological anthropological articulation of

magis requiret, neuter tamen quicquam certi comprehendet, aut tradet nisi divinitus illi revelatum fuerit."

200 Wrightsman, "Andreas Osiander's Contribution," 228.

201 Barker and Goldstein, "Realism," 251.

202 Shank, "Regiomontanus," 202.

203 Wrightsman, "Andreas Osiander's Contribution," 228.

the issue of human knowledge. Consequently, Copernicus's pupil can be reintegrated within the contemporary Wittenberg context.

Based on their shared moderate skepticism about the attainability of natural knowledge, Nicholas Jardine has linked the anonymous preface of Osiander to Melanchthon's attitude. With Osiander, this was proven by his above cited remark about revelation being the only source of certainty. For Melanchthon, meanwhile, his skepticism is attested by frequent references scattered throughout his writings to the meagerness of knowledge about the world. This skepticism did not result in a radical skeptical claim that no knowledge was attainable about the world. Adam, moreover, had been given a basic knowledge of the world and humanity that he passed on to Abel and his descendants.[204] Jardine found Osiander and Melanchthon's epistemology to be representative of a widespread attitude at Protestant universities in the Holy Roman Empire.[205] This moderate skepticism linked Rheticus with Osiander and Melanchthon precisely in terms of their epistemological position on the knowability of the physical world. Melanchton's rebuttal of terrestrial mobility as a physical claim equally has to be framed against this skeptical background. Whereas Rheticus argued for the Bible's obscurity in this matter, Melanchthon found Scripture's teachings to be much more unequivocal on this subject. Melanchthon dealt with Copernicus's position on terrestrial mobility in the *Initia physicae doctrinae* in the entry *Quis est motus mundi?*[206] The composition of the *Initia physicae doctrinae* was in fact a joint effort by three Wittenberg professors: Melanchthon, Paul Eber, and Erasmus Reinhold. The manuscript version shows that Melanchthon and Eber co-authored the textbook, the latter working, *inter alia*, on the order of the cosmos and astronomy, the former writing down the refutation of terrestrial motion. Erasmus Reinhold added annotations and made corrections on mathematical and astronomical details.[207] Unlike Osiander, Melanchthon did aver that Copernicus's hypotheses represented a physical claim about the constitution of the world. He labelled the hypotheses absurd in the first edition of the *Initia* of 1549, seemingly intent on throwing the liberal arts into some confusion.[208] The *praeceptor Germaniae* established what was to become a canonical list of physical, Scriptural, and observational arguments

204 Lerner, Segonds, and Verdet, *Nicolas Copernic*, 1: 473.

205 Jardine, "Forging of Modern Realism," 146–153.

206 Melanchthon, "Initia," 13: 216–220.

207 Omodeo and Regier, "Wittenberg Reception of Copernicus," 88–89.

208 In subsequent editions, Melanchthon's judgment on Copernicus would become less harsh. His opposition to the latter's hypotheses remained unchanged, however; see: ibid., 85–88.

in the debate about terrestrial mobility (against a possible motion of the earth, against the denial of the respective motions of the eighth sphere and the sun, and against the position that the earth is located amidst the stars).[209]

He begins his discussion with the argument of direct visual observation as a sensory reassurance that heaven indeed revolves around the earth.

> And the eyes are witnesses that heaven is carried around in a circle within twenty-four hours.[210]

It is his understanding of the Bible, however, that shows his epistemological kinship with Rheticus. Kenneth Howell convincingly argues that Melanchthon's introduction of Biblical arguments into the discussion was based not on a naïve literalism but on a belief in the perspicuity of Holy Writ.[211] Unlike Rheticus, Melanchthon did find clarity in Scriptural teaching on the subject of the immobility and the centrality of the earth's position. Both, however, admit that only revelation could illuminate the obscurity of the mind. Melanchthon prepared his hermeneutical discussion of terrestrial immobility by emphasizing in two ways the importance of acquiescence to divine testimony. First, he stressed that a good mind embraced divinely revealed truth, and expressed his gratitude to God for having kindled the light that in human minds provided access to God, ruminating on how the knowledge of truth aided and governed this life. The second and more important reason for the chapter in question was the intellectual condition of man:

> Although some natural philosophers ridicule those who appeal to divine testimony, we think that an honorable philosopher, given the great obscurity of the human mind, collects the heavenly statements, and consults divine authority, wherever we can.[212]

His use of the word *caligo* is telling in this respect. Literally, it means fog or darkness but, in a figurative manner, the term indicates the inability to perceive something with the mind. Melanchthon deemed that the human mind

---

209 Barker, "Lutheran Contribution," 48.

210 Melanchthon, "Initia," 13: 216: "Et oculi sunt testes, coelum circumagi viginti quatuor horis."

211 Howell, *God's Two Books*, 53.

212 Melanchthon, "Initia," 13: 216: "Quanquam autem rident aliqui physicum testimonia divina citantem, tamen nos honestum esse censemus philosophiam conferre ad coelestia dicta, et in tanta caligine humanae mentis autoritatem divinam consulere, ubicumque possumus."

had failed to achieve such knowledge by itself and therefore required assistance from divine authority. Like Osiander, he indicated that this matter surpassed the human intellect.[213]

The use of Holy Scripture to rebut terrestrial mobility appears not to have been a standard practice. The four Scriptural passages used by Melanchthon to support terrestrial immobility, the central position of the earth in the universe, and solar mobility will be seen to recur throughout hermeneutical discussions during the decades that followed. The passages on the motion of the sun, were as follows: Psalm 19:4–6 ("In them hath he set a tabernacle for the sun, Which is as a bridegroom coming out of his chamber, and rejoiceth as a strong man to run a race. His going forth is from the end of the heaven, and his circuit unto the ends of it: and there is nothing hid from the heat thereof"); on the stability of the earth, Psalm 104:5 ("Who laid the foundations of the earth, that it should not be removed for ever"); and Ecclesiastes 1:4–5 ("One generation passeth away, and another generation cometh: but the earth abideth. The sun also ariseth, and the sun goeth down, and hasteth to his place where he arose") which confirms that "the earth remains for eternity, the sun rises and goes down, and hastens to its place where it rises." Finally, Melanchthon mentioned the miraculous standstill of the sun at Joshua 10:12 ("Then spake Joshua to the Lord in the day when the Lord delivered up the Amorites before the children of Israel, and he said in the sight of Israel, Sun, stand thou still upon Gibeon; and thou, Moon, in the valley of Ajalon"); although he does not give the verse, nor any more details than that God had made the sun stand still and return, it is clear that he had this passage in mind from later uses of the same verse in similar contexts. The story goes that, facing defeat in war, Joshua prayed to God to make the sun and the moon stand still so that the Israelites could still win the battle. The standstill of the sun was the divine response to Joshua's prayer. Concluding his discussion of the Biblical arguments, Melanchthon affirmed that man had to delight in the truth that was revealed by divine testimony.[214]

In spite of his additional physical and observational arguments, his reference to the obscurity of the human mind shows that the criterion of proof, in his opinion, rested with divine authority: if a *propter quid* of the celestial phenomena was to be found, it was provided by God's Word.[215] At the time, there

---

213 Blumenberg, *Genesis*, 327.

214 Melanchthon, "Initia," 13: 217.

215 Melanchthon's physical arguments were as follows: The first is a syllogism that explains that because the earth is a single simple body, it displays a single motion and cannot be moved by means of a circular motion. The second runs that the center is necessarily immobile for all that is heavy tends towards it to rest. Thirdly, all parts of the earth, and all mixed

were no compelling physical arguments in favor of terrestrial immobility.[216] Hence, the physical arguments were adduced because they were in agreement with Holy Writ. On Melanchthon's appraisal of the motion of the earth, "The Bible," in the words of Methuen, "was the supreme arbiter of truth."[217]

As in Rheticus's discourse in the *Epistola de terrae motu*, Melanchthon's views on the intellectual condition of obscurity and his recourse to Scripture as the sole source of truth issued from an epistemology built on the deficient postlapsarian condition of man. He too formulated the problem of knowledge in theological anthropological terms and explained man's lack of knowledge as resulting from his temporal position between the Fall and afterlife. In the prefatory letter of the *Initia*, Melanchthon explained to his correspondent Meyenburg that the divine wisdom responsible for creating the world remained inscrutable to human beings while on earth.[218] Divine wisdom and power were rationally unfathomable. The cause of this earthly condition was the Fall. In his preface to his commentary *De anima* (1540), Melanchthon blamed the devil himself for the severe diminution of the light of the soul.

> If the soul had kept that light and harmony which are bestowed upon it in its creation by divine providence there would be less need for other learned men, and it would examine its nature by its own sharpness of vision. However, now that—like some outstanding picture by Apelles bespattered with mud—it lies in the body, buried in hideous darkness, there is the greatest need of knowledge that should bring it forth and put it in our view, and show how such a great wound the enemy inflicted on it, who overthrew the first ancestor of humankind, and show also the traces of the divine image on it, and the remains of the heavenly gifts.[219]

beings that dominate the earth, tend downwards to the surface of the earth to rest. Thus, the earth rests in the universe. Melanchthon concludes by means of physical conjecture: the earth is "a womb" of many living creatures and that, if moved at such a speed, all things would disperse and could not be kept together; see: ibid., 13: 218–219.

216 Howell, *God's Two Books*, 57.

217 Methuen, "On the Threshold," 674.

218 Melanchthon, "M. Meienburgo," 7: 476.

219 Melanchthon, "Preface to the *Book on the Soul*," 146; Melanchthon, "Hier. Baumgartnero," 3: 909: "[…] si retuinisset eam lucem ac harmoniam, quae in conditione ei divinitus attributa est, minus esset opus aliis doctoribus, suam ipsa naturam sua acie introspexisset. Nunc vero postquam velut excellens aliqua, Apellis tabula coeno conspersa, iacet in corpore, tetra caligine obruta, maxime opus est doctrina, quae proferat eam, et in conspectum ponat, ac ostendat quantum ei vulnus fecerit hostis, qui primum parentem generis humani evertit; monstret etiam vestigia divinae imaginis in ea, et reliquias coelestium donorum."

Due to original sin, temporal natural philosophy did not represent a full comprehension of nature: As Melanchthon wrote in the *Chronicon Carionis*, Adam had understood many things about heaven and earth before the Fall, that human minds afterwards could no longer grasp. He passed on to Abel only the seeds of the doctrines about God, the celestial motions, and living creatures.[220] While on earth, man could gain cognition only of these seeds of an integral natural philosophy.[221] In his preface to *Initia*, Melanchthon admitted that, in this life, the obscurity of the mind prevailed as far as causes of the divine works and the inmost part of nature were concerned.

> Even though nature cannot be investigated into the inmost part, nor the causes of such wonderful works be understood until we hear before us the deliberation of the Father, the eternal architect himself, the Son and the Holy Spirit, nonetheless we must undertake, despite our obscurity, an examination and consideration of the order of beautiful bodies in the world, for the sake both of the knowledge of God, and of virtue, that is, of loving and preserving order and measure in all actions.[222]

Once again, notice the use of the word *caligo*. With respect to the invisible causes of natural phenomena and the inmost parts of nature (causality and substance), the philosopher gazed into obscurity. The *praeceptor* added in the following phrase that he acknowledged his own infirmity.[223] Exhaustive knowledge of nature was to be obtained only after death, when abiding in the company of God. Melanchthon pictured heaven as an eternal place of learning. In the eternal academy, he wrote, we will learn to know the undeteriorated natural philosophy, where the architect of the world will himself show us the ideal of the world itself.[224]

> And there is another eternal one [place of learning], where God will be all for all, and will impart His wisdom and justice to us, where we shall

---

220 *Chronicon Carionis*, 15.

221 Melanchthon, "Baumgartnero, filio," 7: 1125–1126.

222 Melanchthon, "M. Meienburgo," 7: 473: "Etsi autem natura rerum non potest penitus perspici, nec caussae tam mirabilium operum prius intelligi poterunt, quam ipsius architecti aeterni patris, filii et spiritus sancti deliberationem coram audiemus, tamen nunc quoque in hac caligine nostra qualiscunque aspectio et consideratio ordinis pulcherrimorum in mundo corporum aditus est et ad agnitionem dei, et ad virtutem, hoc est, ad amandum et conservandum ordinem et modum in omnibus actionibus."

223 Ibid., 7: 475.

224 Ibid., 7: 473.

> see Him in person, and in the Son, the Word, we shall behold not only the ideas and causes of the workings of the world, but also the wonderful joining of divine and human nature, and the plan for the restoration of mankind. There we shall behold the Maker and contemplate the causes of things, and we shall have complete insight into nature.[225]

The face-to-face vision of God would elicit an exhaustive cognition of nature; divine wisdom would be fully understood. In between the Fall and the afterlife, man had to make use of Holy Writ. There was certainty to be obtained in the postlapsarian condition of man but this was limited to the experiential level. In the *Initia*, Melanchthon concluded the passage about whether there is certainty in physical doctrine by underlining an *a posteriori* perspective on certainty and, in so doing, ruled out certainty regarding the *a priori* in natural philosophy: "Let us no longer seek demonstrations of why fire burns, but content ourselves by experience to hold that it is certain."[226] The lack of knowledge about the meta-visible aspects of nature (or rather visible creation) aligns with Melanchthon's perception on the inability of humans to attain theological truth by their own capacity.[227]

As a result of the Fall, human beings by their own rational capacity could achieve exhaustive knowledge about God only when encountering Him face-to-face in the afterlife. Knowledge of God, spirits, and substances was tenuous and to be achieved by reasoning from effects to causes. Knowledge of these matters could only be acquired *a posteriori*, by reasoning from effects to causes.[228] Melanchthon had modeled his approach of knowledge in natural philosophy and theology in accordance with the Lutheran principle of distrust of human reason. Metaphysical or *a priori* knowledge (invisible causes), either in theology or in natural philosophy, was not deemed attainable by postlapsarian reason, unless it was informed by revelation. In Peterson's opinion, Melanchthon forestalled "the possibility of dogmatic philosophical

225 "Preface to the *Book on the Soul*," 153; Melanchthon, "Baumgartnero, filio," 7: 1124: "[…] aeternam alteram, in qua Deus erit omnia in omnibus, et suam nobis sapientiam et iustitiam communicabit, ubi coram ipsum cernemus, et in filio conspiciemus non solum ideas et caussas opificii mundi, sed etiam admirandam copulationem divinae naturae et humanae, et consilium restitutionis generis humani. Ibi conditorem intuentes et caussas rerum considerantes, naturam penitus."

226 Melanchthon, "Initia," 13:189: "Nec longius quaeramus demonstrationes, cur ignis urat, sed contenti simus experientia, quam ducamus esse certam."

227 Kusukawa, *Transformation of Natural Philosophy*, 164.

228 Eadem, "Introduction," xxi–xxii.

apriorism for humans in our present state."[229] When speaking of Melanchthon as an exponent of Protestant Scholasticism, it is important to differentiate his view on the capabilities of human reason from those of the Scholastic theologians and natural philosophers.

Melanchthon's mistrust of novelty, which he identified in Copernicus's heliocentric model of the universe, ought to be understood against this theological anthropological background. In his opinion, reviving the paradox of Aristarchus of Samos was absurd, and was rather a means to display one's ingenuity or love of novelty, a counterargument that both Rheticus and Copernicus had tried to anticipate.[230] Such "absurd opinions" set a bad example, in his opinion.[231] In my view, Melanchthon underlined the importance of proceeding cautiously in the study of nature because of the obscurity of the mind and because of the association between novelty and heresy: this is the reason why he relied on ancient authoritative opinions which had proven their trustworthiness over the centuries. Earlier we observed his testing of the philosophical legacy of the Epicureans, Peripatetics, Academics, and Stoics against true doctrine, embodying the one eternal truth. He preferred Aristotle because, out of the four ancient schools, this was the least erroneous or absurd in relation to Scripture.[232] Melanchthon recommended his students to follow the authority of tradition.[233]

The argument of authority that Melanchthon introduced in his rebuttal of the positing of the advancing of terrestrial mobility by the advancing of physical reasons was based on an assessment of the weakness of the abilities of the human mind: the authorities to whom he turned were selected on the basis of the extent to which they were in agreement with the clear teaching of Scripture.

229 Peterson, "Philipp Melanchthon," 69–75.
230 Melanchthon, "Initia," 13: 216.
231 Ibid.
232 Melanchthon, "Erotemata dialectices," 13: 655–658.
233 Lerner, Segonds, and Verdet, *Nicolas Copernic*, 1: 474.

CHAPTER 5

Tycho Brahe's Turn to the Book of Nature

Allusions to the Adamic figure in cosmological debate did not stop at the geographical borders of the sixteenth-century Holy Roman Empire. Danish astronomer Tycho Brahe too indicated how hard it was for "the corrupted nature of the earthly Adam" to follow the path of truth.[1] The Lord of Uraniborg aspired an empirical reform of astronomy, and believed that future observations would become superfluous due to the accuracy of his own observational data.[2] In his opinion, his observations allowed him to act rightfully as an astronomer-philosopher and showed that there was no reason to limit divine providence to the teachings of the Scholastic or Peripatetic philosophers.

Observations of a new star that had appeared in the constellation of Cassiopeia in 1572 and a new comet in 1577 made Tycho Brahe—and a select few other astronomers—step away from particular tenets and the methodology of Scholastic natural philosophy. Cometary research had convinced him of the illegitimacy of several canonical tenets regarding the celestial realm: the incorruptible nature of the heavenly realm, the existence of solid celestial spheres, and the sublunary trajectories of comets. According to Aristotelian cosmology, the realm above the moon consisted of an ethereal substance known as quintessence. This celestial matter was characterized by immutability which manifested itself, among other things, by the ever circular and uniform motions of the celestial bodies.[3] Comets had previously been defined as sublunary phenomena.[4] Given the widespread belief in the immutability of the celestial

1 Brahe, "Astronomiae instauratae progymnasmata," 2: 316: "[…] natura Adami terrestris corruptae […]."

2 Daston, "Empire of Observation," 94; Mosley, *Bearing the Heavens*, 27.

3 Grant, *Planets*, 190–192.

4 Ibid., 353–354. According to Edward Grant, the hardness or softness of the solid celestial spheres was a non-issue in medieval natural philosophical commentaries and was actually a theme absent from the writings of Aristotle, who assumed the corporeality of the celestial orbs. Grant conjectures that the persistent association between solidity and hardness of the celestial spheres arose either between the publication of *De revolutionibus* in 1543 and the appearance of the new star in 1572, or after Tycho's publication of *De mundi aetherei recentioribus phaenomenis* in 1588. For further information, see: Grant, *Planets*, 324–326, 345. On the subject of Tycho's role in the late sixteenth-century dispersal of the celestial spheres, Adam Mosley has argued that Tycho was not said to have offered a conclusive disproof of the existence of celestial orbs. Michel-Pierre Lerner argued that this was a joint operation in which Tycho was only one of the participants; so too did Bernard Goldstein and Peter Barker

 DOI:10.1163/9789004387751_007

realm, the notion of a celestial novelty was somewhat of an oxymoron up to the seventeenth century. Much effort was put into realigning celestial novelties with the predetermined regularity of celestial motion,[5] a strategy that Tycho loudly and decisively disavowed. Amongst his tactics of persuasion were the systematic refutation of other views, arguments for the superiority of mathematical certainty to philosophical opinion, and his arguments for demonstration from precise observation. Other astronomers who were in the same process of walking away from certain canonical tenets based on observation, such as Christoph Rothmann, employed similar elements in rhetorical strategy.[6]

The Lord of Uraniborg overtly reshuffled the epistemological cards, and did this outside of a university context. A clear pattern is discernable between Tycho and fellow Renaissance mathematicians Rheticus and Regiomontanus: The Scholastic hierarchy of sciences was breached to the detriment of natural philosophy and became paired with a discrediting of Scholastic natural philosophy. In his self-published writings, printed by means of a privately owned printing press outside the context of a university, the Danish astronomer likewise dispensed with the Scholastic appraisal of the subordinate role of the mathematical disciplines to natural philosophy.[7] His criticism against the Peripatetics was funneled into a vehement polemic against the natural philosophical instruction provided at academies. Similarly to Rheticus, these disciplinary incursions were articulated in theological anthropological terms. The Renaissance theme of opinion versus apodictic certainty takes a theological anthropological turn, delivering an argument about the deceitful human mind and the asymmetrical relationship between man and God. Tycho articulated his preference for observation over Peripatetic philosophy by means of the Renaissance metaphor of the Book of Nature. The metaphor had become ubiquitous in early modern scientific discourse. It implies that nature was an alternative text, a book authored by God Himself, being a locus of revelation in which the visible effects of divine providence were put on display before human readers.[8] According to Peter Harrison, the evolution of the medieval approach to

---

by highlighting the role of the astronomer in Kassel, Christoph Rothmann in this respect; see: Goldstein and Barker, "Role of Rothmann," 385–403; Lerner, *Le Monde des Sphères. II*, 58–66; Mosley, *Bearing the Heavens*, 72–73; Granada, "Did Tycho," 125–145.

5 Boner and Tessicini, "Introduction," VII.

6 Jardine, "How to Present," 159–184.

7 Westman, *Copernican Question*, 230.

8 Hacking, *Emergence of Probability*, 30; Harrison, *Bible*, 45. The Book of Nature is becoming an increasingly studied historiographical theme for the early modern period. For further reading, see: Jorink, *Reading the Book of Nature in the Dutch Golden Age 1575–1715*; a range of articles

nature as a mirror to the early modern description of nature as a book implied a newly acquired independence for the realm of nature.[9] Tycho's turn to the Book of Nature was motivated by a pessimistic interpretation of the epistemic abilities of postlapsarian man, coupled with an imperative to proceed piously while investigating the heavens. His writings explicitly show that distrust of human reason and intellectual piety were essentially two sides of the same Lutheran epistemological coin. Applying a Lutheran anthropological perspective to Tycho's disciplinary incursions provides answers to certain questions which have either been sidestepped or left unanswered in much modern historiography.

1 The Wittenberg Connection

Born into the high nobility of Denmark, Tyge Brahe Ottesen was educated in a "Lutheran Wittenberg spirit" wherein he also pursued his astronomical activities. At the age of twelve, he commenced his studies at the University of Copenhagen which had a strong Wittenbergian orientation. In imitation of the teachings of Philipp Melanchthon, the Copenhagen curriculum had a pronounced Aristotelian and Scholastic slant.[10] From a Reformed point of view, the Danish university was a Philippist bastion. The theologians present at Copenhagen were almost exclusively disciples of Melanchthon and had pursued at least a part of their studies at Wittenberg University.[11] Besides prospective theologians, many Danish scholars had obtained degrees at Wittenberg, to such an extent that Wittenberg University has been labeled the postgraduate school of the University of Copenhagen.[12] Tycho Brahe's astronomy, and his astrology in particular, have been described as having a distinct Melanchthonian religious dimension in terms of his Lutheran outlook on the providential rule of the universe.[13]

---

in both volumes of *Nature and Scripture in the Abrahamic Religions*, edited by van der Meer and Mandelbrote; van Berkel and Vanderjagt, *Book of Nature*.

9 Harrison, "'Book of Nature'," 1–26.

10 For further information about Tycho's years at the University of Copenhagen in the sixteenth century, see: Christianson, "Tycho Brahe," 198–203.

11 Christianson, "Tycho Brahe's German Treatises," 114.

12 This reference is taken from Victor Thoren's biography of Tycho Brahe. Unfortunately, Thoren does not identify which modern scholar coined the expression; see: Thoren, *Lord of Uraniborg*, 22.

13 Granada, "Tycho Brahe, Caspar Peucer," 567; Westman, *Copernican Question*, 92.

After his studies, Tycho did not receive the usual education for young noblemen preparing them for military or civil service to the Danish king.[14] In 1562 the young Tycho instead set off on a grand tour across multiple foreign universities, a tour that lasted for eight years. During this period, he visited the University of Wittenberg on several occasions. The longest of these stays was from April until September 1566. For unknown reasons he did not chose to matriculate there. Over a period of eight years he studied at the Lutheran universities of Leipzig (1562–1565) and Rostock (1566–1567). He also visited the universities of Wittenberg, Basel, and Ingolstadt.[15] Throughout his life he paid several visits to Wittenberg and showed himself to be well informed about the astronomical tradition at Wittenberg University. Though he never met Rheticus or Reinhold in person—the latter died when Tycho was a child, while the former had left Wittenberg in 1542—, he wrote about Reinhold's work as a mathematician in very glowing terms and mentioned having read Rheticus's *Narratio prima*.[16] According to Karl Burmeister, he attempted to acquire Rheticus's library.[17] Beyond having once met at Wittenberg University in 1562,[18] Tycho and Caspar Peucer briefly corresponded on the former's geoheliocentric model and other cosmological themes in the late 1580s.

Having started a family with Kirsten Jørgensdatter in the autumn of 1574, the Danish nobleman delivered a series of lectures on planetary theory at the University of Copenhagen. Three years later, Tycho declined an offer to become rector of the university.[19] Following another period of travel across astronomical centers in Central Europe, among them the Kassel observatory, in 1576 the Danish King Frederick II (1534–1588) granted Tycho the isle of Hven as a fiefdom for life. As royal governor he founded an observatory, an alchemical laboratory, an academy, and a court.[20] The first stone of the castle observatory, devoted to the muse of astronomy Urania, was laid on the island in August that year. Uraniborg was organized as a scientific observatory that housed both

14 This biographical introduction to Tycho relies mainly on the biography written by Victor Thoren. Another important biographer was John Louis Emil Dreyer, see: Dreyer, *Tycho Brahe*.

15 For an elaborate description of Tycho's education, see: Thoren, *Lord of Uraniborg*, 8–37.

16 For examples of Tycho's praise of Reinhold, see: Brahe, "Astronomiae instauratae progymnasmata," 3: 146, 150; and Tycho Brahe, letter to Caspar Peucer, Uraniborg, 13 September 1588, 6: 137. For Tycho's reference to having read Rheticus's *Narratio prima*, see: Brahe, "Astronomiae instauratae progymnasmata," 2: 31.

17 Burmeister, *Georg Joachim Rhetikus*, 1: 181.

18 Westman, "Three Responses," 306.

19 Idem, "Astronomer's Role" 123.

20 Omodeo, *Copernicus*, 45.

Tycho's family as well as a community of assistants, the so-called *familia*.[21] Tycho's time at Uraniborg came to represent the height of his lifelong regime of intensive observational activity: over a period of more than twenty years, he observed the sky every clear night.[22] An additional basement observatory named Stellaeburg (Stjerneborg) was constructed in 1584.[23] In the early 1580s, a printing house was build and a printing press installed, allowing Tycho to self-publish his own works and correspondence. His *De mundi aetherei recentioribus phaenomenis* (1588) was published from this press, as was the *Epistolarum astronomicarum liber primus* (1596). In addition, he engaged in an international epistolary network of mathematicians and other learned men, making him a member of the virtual Republic of Letters.[24] The Danish astronomer was keen to spread and discuss his work, and Hven was visited by astronomers and aristocrats from all over Northern Europe.[25] This state of affairs continued until 1596, when a new king ascended to the Danish throne. For reasons which remain unclear, Christian IV (1577–1648) would no longer grant the astronomer his annual pension. Tycho could thus no longer maintain his expensive lifestyle at Hven. On 15 March 1597, the last observation was made from Uraniborg castle. In 1599, Tycho emigrated to the Kingdom of Bohemia. He had won the favor of Emperor Rudolf II (1552–1612) who promised to build him a second Uraniborg in the surroundings of Prague. However, Tycho died of illness in the fall of 1601 and was buried in the Teyn Cathedral of Prague. Kepler finished and published Tycho's posthumous *Astronomiae instauratae progymnasmata* in 1602.

2 Into the Wrestling Arena

Tycho's empirical reform of astronomy was built on thorough observation from the platforms added to the castle of Uraniborg, carried out with meticulously assembled instruments, which featured in illustrations in his *Astronomiae instauratae mechanica* (1598). The astronomer claimed to be breaking new ground when it came to observing heaven: the quality of his highly expensive instruments was determined by their enormous size, material, and partic-

21 Christianson, *On Tycho's Island*.

22 Daston, "Empire of Observation," 94.

23 Thoren, *Lord of Uraniborg*, 180.

24 Mosley, *Bearing the Heavens*, 33–38; a schematic outline of Tycho's network of correspondents is found at ibid., 36.

25 Omodeo, *Copernicus*, 45.

ular mode of construction. The instruments were depicted as metallic incarnations of geometrical techniques; the instruments and machines were elaborated "perfectly according to geometry" and their construction was in keeping with the art of mechanics.[26] Tycho stated that his perfected instruments were "exempt from all error."[27] It was celestial observation, made with delicate instruments, "not guilty of any imperfection," that changed his mind regarding the reality of solid celestial spheres and the positions of comets as sublunary meteors.[28] These instruments had established facts for the positions and motion of the novas and the comets. Tycho often lamented the instruments employed by other contemporary astronomers and indicated the concomitant failings as the source of many of their wrongful suppositions. The nobleman argued that his method allowed the Book of Nature—the divine truth itself—to speak, and to do so instead of the deceitful human mind. This is precisely the crux of his harsh critique of the Peripatetic philosophers who, in his opinion, let Aristotle's writings shape their findings on the new phenomenon in heaven. One had to have an open mind when studying divine creation, which means not allowing the interpretation of one's observational results to be predetermined by authoritative books. The Lord of Uraniborg thus delivered a most explicit and vehement repudiation of the Peripatetic philosophers. Given the prevalence of Aristotle's teaching in the academic curriculum of the time, Tycho's decision to take issue with Peripatetic natural philosophy was not a minor task. In the words of Tycho's son-in-law—the editor of the second edition of *De mundi aetherei recentioribus phaenomenis*, Franz Tengnagel van Kamp (1576–1622)—the astronomer prepared himself for combat and descended into "the wrestling arena" to discuss his findings on the new star and the comet of 1577.[29]

2.1 *Certainty vs. Authority*

Tycho Brahe used the argument of unambiguous certainty, obtained by scientific observation and mathematical demonstration, to vindicate his rejection of the canonized opinion on the incorruptibility of the celestial realm and the existence of solid celestial spheres. His complementary denunciation of the Peripatetics may be readily reduced to a dispute between the argu-

26 Brahe, "Astronomiae instauratae progymnasmata," 2: 330.

27 For one of many examples, see: ibid., 2: 281.

28 Ibid., 3: 112.

29 This was a prefatory letter included by Franz Tengnagel in the 1603 edition of *De phaenomenis*. Tengnagel was married to Tycho's daughter Elizabeth in 1601; see: Thoren, *Lord of Uraniborg*, 463; Brahe, "De mundi aetherei recentioribus phaenomenis," vol. 4.

ment of authoritative opinion and the argument of demonstrative certainty. His demonstrations had proven the Peripatetic opinion wrong insofar as the comets he observed were confirmed to be travelling above the moon, and the new star was proven moreover to be located among the fixed stars.

For Tycho, authority implied uncertainty and doubt. The contrast between a demonstration-based certainty and conjecture-based uncertainty emerges many times in his books on the restoration of astronomy.

> [...] it were better to have concluded, against long-received opinion, due to observation and certain demonstration, that heaven sometimes receives new bodies, and is even susceptible to mutation and dissolution, than to contradict certain observations and geometrical truth on account of the uncertain and dubious conjectures of natural philosophers, supported by the authority of Aristotle alone.[30]

Instead of interpreting observational results by means of authoritative opinions, the observational astronomer suggested a procedure of induction in which observations were to serve as the principles of geometric and arithmetic demonstrations. He believed that this way of proceeding would render the Scholastic opposition "dumbfounded" and "speechless."[31] He reproached the Peripatetics and the other philosophers for their reliance on an authoritative opinion, the certainty of which had not been proven:

> They draw this commonly held opinion principally from the Aristotelian school, and, trusting in his authority, accept plausible arguments devised by him, which no experience has ever proven, no observation from careful Instruments has buttressed, and no mathematical demonstration of figures [geometry] and of numbers [arithmetic] has confirmed.[32]

---

30 Brahe, "Astronomiae instauratae progymnasmata," 3: 55: "[...] cum observatione atque demonstratione indubia standum, et hinc coelum aliqua nova recipere corpora, etiam mutatione et dissolutioni obnoxia, contra diu receptam opinionem, concludendum fuisset, quam quod ob incertas et dubias physicorum coniectationes solius Aristotelis authoritate fulcitas, observationibus ratis et geometricae Veritati contradicendum."

31 Ibid., 2: 382.

32 Brahe, "De mundi aetherei recentioribus phaenomenis," 4: 7: "Hanc opinionem vulgo receptam inprimis ex schola Aristotelis hauserunt, eiusque autoritate confisi plausibilibus ab eo excogitatis argumentis acquieuerunt, quae tamen nulla unquam comprobavit experientia, nulla adinvenit organis exquisitis facta observatio, nulla stabilivit mathematica figurarum numerorumque demonstratio."

The Aristotelian natural philosophers were depicted as proceeding by blind faith and as drawing untrustworthy, pagan conclusions. The philosophers, Tycho argued, had behaved in a godless and idolatrous manner by wrongfully attributing genuine piety to the Stagirite, like a demi-god.[33] Likewise, Tycho invoked Aristotle's paganism to undermine the traditionally subservient status of mathematical astronomy: if the axioms of Aristotle's philosophy were true, experience would confirm them—but it did not. Furthermore, as a heathen, Aristotle was not infused with divine inspiration. Any suggestion that the pagan philosophers were divinely inspired, and that they had solved all issues about nature, was not only absurd but also impious.[34]

Observation and demonstration had proved that the new star had emerged among the fixed stars in the upper region of heaven. Tycho blamed the Peripatetics for having produced centuries' worth of bias based on "authority" and subtle "plausibilities" regarding the position and generation of comets below the moon.[35] He opposed the intellectual strategy of letting preconceived Peripatetic judgment shape a biased interpretation of observational results. One example was Regiomontanus, to whom he elsewhere refers approvingly, and who had interpreted his cometary parallactic calculations in accordance with Aristotelian philosophy:

> However, he [Regiomontanus] agrees on the whole with Aristotle about the elementary position of comets, for his authority had occupied all chairs and all access to the sciences so that they judged it impious to deviate a hair's breadth from what he approved; to such an extent that the free path of truth was always closed to the servile minds of men.[36]

According to the Danish astronomer, common Scholastic procedures at universities, fleshed out in an uncritical adherence to Aristotle's philosophy, had barred the free path of truth: his authority had rendered incredible the possibility of a supralunary localization for comets. University instruction came close to being represented as a form of indoctrination, effecting an intellectual attitude of servile obedience that prevented the mind from accessing truth.

---

33 Brahe, "Astronomiae instauratae progymnasmata," 3: 142.

34 Ibid., 3: 153.

35 Brahe, "De mundi aetherei recentioribus phaenomenis," 4: 10.

36 Ibid., 4: 124: "Licet vero cum Aristotele de cometarum elementari situ prorsus consentiret; Ipsius enim authoritas omnia pulpita, omnes scientarum aditus occupaverat, ut nefas iudicarent, ab eius placitis latum unguem discedere; Adeo semper praeclusa est, servilibus hominum ingeniis libera veritatis via."

University once more was clearly not perceived as a place for free philosophical debate but rather as a place for canonical confirmation. Whereas Rheticus still argued for compatibility with ancient authority on the levels of methodology and purpose, the Danish astronomer did not hint at such conformity. In the *Narratio prima*, the former had argued that although the outcome was different, Copernicus had aimed his arrows by the same method at the same target as Ptolemy—namely, he had intended to provide explanatory hypotheses for the observed celestial effects. Rheticus's vindication of Copernicus's right to theorize, and Tycho's plea for his right to assert deviating philosophical positions both rely on resisting a theory-laden observation of the celestial phenomena. To proceed in such a manner was the result of human weakness.

More than once in his *Progymnasmata* and *De phaenomenis*, Scholastic opinions and methodology were explicitly associated with a human condition marked by shadows, fallibility, and the tendency to deviate from the path of truth. The second part of the tenth chapter of *De phaenomenis* deals with four men who upheld the Peripatetic opinion on comets as sublunary instead of supralunary phenomena. Tycho argued that the "more certain rule of observations and demonstrations" revealed "how easy the way of error is, and how easily mortals are taken away by devious things, being on an abstruse footpath of truth."[37] The Peripatetic opinion was characterized as the mortal path of error that was both the easiest as well as the most distant from the path of truth. Another instance concerns an extensive discussion of the cometary research of Thaddaeus Hagecius ab Hayek (1525–1600), astronomer and court physician to Emperor Rudolf II. The Bohemian astronomer agreed with Tycho on the position of the new star in the high ether but held that comets could be created in both the sub- and supralunary realm.[38] Tycho extolled Thaddaeus for his noble love of truth, but again blamed his feeble human nature for having diverted him from the path of truth.[39] The "mortal errors" to which Tycho believed the academic philosophers exposed themselves contrast sharply with the eternal certitude he claimed to have acquired from his scrutiny of the celestial phenomena.[40] Urania, Franz Tengnagel wrote in an ode to his father-in-law, "vanquished the shadows."[41]

37 Ibid., 4: 259: "[…] easque ad observationum et demonstrationum certiorem normam diligenter ponderabo […] quam procliuis sit errorum via, et quam facile mortales in devia abducantur, adeo ut veritatis abstrusum tramitem […]."

38 Brahe, "Astronomiae instauratae progymnasmata," 3: 26–27.

39 Idem, "De mundi aetherei recentioribus phaenomenis," 4: 294–295.

40 Ibid., 4: 51.

41 Brahe, "Astronomiae instauratae mechanica," 5: 93.

Despite the feeble condition of human nature, Tycho did not admit to a radically skeptical position. Mathematical demonstrations emerged as the stronghold of certitude, a light for the feeble human mind. Mathematical demonstration was followed by geometrical demonstration as a second support for Tycho's method of determining the new star's position. In light of this, it is interesting to note how he adduced geometrical demonstration as a counterargument to a radical skepticism, because demonstration suspended the possibility of contradiction.[42] Geometrical demonstration itself was the proof of the existence of certainty. Given the fact that Tycho considers all arts to be divine gifts,[43] this line of reasoning is identical to Melanchthon's position. In the *Erotemata dialectices*, the latter stated that the pure mathematical disciplines were a divine gift handed to humanity in order to achieve firm and unwavering certainty. He used this argument against the Academic stance that all things were uncertain.[44] Tycho was arguing in favor of irrefutable demonstrations that left no room for hesitation or contradiction, for proceeding "without prejudice" and the desire to know and embrace what is true.[45] He called mathematicians "lovers of truth" which emphasizes the mathematicians' interest in knowing and embracing what is true based on invincible arguments.[46] A survey of his use of the term demonstration (*apodixis*, *demonstratio*) throughout his writings makes clear that the word was almost always, as a rule, equipped with adjectives like infallible, certain, and irrefutable.[47] The everlasting certitude of which Tycho believed the mathematical disciplines were capable, was represented iconographically as well in some of the hieroglyphs with which the astronomer had his instruments adorned. His iconographical interpretation of the relation between astronomy, geometry, and arithmetic was represented in a hieroglyph on top of the great azimuth semicircle that depicted the three disciplines as virgins. The highest positioned virgin was Urania who was dressed in a white gown with golden asterisks and silver fringes, and represented the art of astronomy. On a lower level, we find her two handmaidens, the virgin of Geometry on the left, and the virgin of Arithmetic on the right. One of the attributes worn by both Arithmetic and Geometry on their heads was a garland of laurel which Tycho described as representing the arts' potential for

42 Idem, "Astronomiae instauratae progymnasmata," 2: 382.

43 Ibid., 3: 153.

44 Melanchthon, "Erotemata dialectices," 13: 657.

45 Brahe, "De mundi aetherei recentioribus phaenomenis," 4: 10.

46 For instance, see: Brahe, "Astronomiae instauratae progymnasmata," 3: 26.

47 For more examples, see: Brahe, "De mundi aetherei recentioribus phaenomenis," 4: 7, 10, 180.

pure and eternal truth.[48] The Danish astronomer extended the scope of this eternal certitude to scientific observation. His observations were described as indubitable.[49] The key to obtaining such infallible results, and for the credibility of the sensible data, were the instruments. The results of mathematical astronomy were intrinsically certain and required no further endorsement by either authority or common opinion.[50] Their trustworthiness was founded on more than just "a good thought come out of human reason."[51] As such, Tycho was a precursor of seventeenth-century physics, where the probable accounts of Aristotelian physics were increasingly downplayed in favor of the mathematical sciences that were able to offer certain demonstration of quantitative relations.[52]

2.2 *"O servitus"*

Wittenberg University was not spared Tycho's criticism of academic astronomy. Reading his accounts of what Michael Weichenhan has called the Wittenberg interpretation of the new star,[53] things seemed not to have changed since Melanchthon's exposition of dialectical rules in the *Erotemata dialectices*, and could readily be identified as instances of a Protestant Scholasticism, wherein maintaining proximity to Aristotelian teaching and preserving the Scholastic hierarchy of sciences were key.

Tycho differentiated three groups of astronomers in reference to the new star of 1572. He himself had found that there was no parallax, meaning that the new star was not subject to motion and stood in a very remote position from the earth.[54] Chapter eight of the *Progymnasmata* discusses the astronomers who likewise had established no parallax for the new star and thus localized the phenomenon in the upper celestial region of the fixed stars. This group was termed "the nullists" by Robert Westman, in reference to the absence of parallax.[55] The chapter which followed dealt with those astronomers, like those at Wittenberg,

---

48 Brahe, "Astronomiae instauratae mechanica," 5: 41–42.

49 For one of many examples, see: Brahe, "Astronomiae instauratae progymnasmata," 2: 382.

50 Brahe, "De mundi aetherei recentioribus phaenomenis," 4: 180.

51 Tycho Brahe, "Vonn der Cometten," 4: 387: "[…] allain aus subtilen argumenten durch die vernunfft […]."

52 Dear, *Discipline*, 3.

53 For an elaborate discussion, see: Weichenhan, *"Ergo perit coelum,"* 472–496.

54 The use of this computational technique has been traced in astronomical writings as ancient as those of Aristarchus but only in the fifteenth century was it applied in cometary research by Regiomontanus; see: Donahue, *Dissolution*, 36; Westman, *Copernican Question*, 231.

55 Westman, *Copernican Question*, 230.

who did situate the new phenomenon in the supralunary realm but not as high as the upper region of the fixed stars, since they had established some parallax for the new star. Chapter ten considers those scholars who did not apply parallactic calculation and did not believe that the star was new but still situated the phenomenon in the aerial or sublunary realm.[56]

The Wittenberg astronomers at the time interpreted the new star as a comet, an instance of special providence. The comet was thought to be located above the moon but was not situated in the remote sphere of the fixed stars, based on the fact that they did establish some parallax for the new phenomenon. Unlike Tycho, they defined it as an unusual comet made of elementary matter. Tycho scrutinized the Wittenberg contributions to the debate through correspondence, excerpts of public debates, and a treatise on the subject which he had been able to study, transcripts of which he provided in his work. The contributors were all professors in mathematics at the University of Wittenberg at certain points in their careers. Besides Caspar Peucer, the other Wittenberg astronomers under scrutiny included Wolfgang Schuler (*d.*1575), Caspar Peucer Jr., and Johannes Praetorius (1537–1616) who were jointly professors in higher mathematics from 1571 until 1577.[57]

From Caspar Peucer's letter to Landgrave of Hessen-Kassel William IV—included in *Progymnasmata*—Tycho deduced two main obstacles to the naming of this new phenomenon as a star. Even though he did initially term the nova a new star, in the end he resiled from this position for reasons of incompatibility with the Bible (notably, Gen.2:2: "And on the seventh day God ended his work which he had made; and he rested on the seventh day from all his work which he had made") and due to incompatibility with Peripatetic opinion.[58] On 5 March 1573, the son of Caspar Peucer, who was of the same name as his father, addressed an academic *quaestio* before a public assembly at Wittenberg University.[59] He asked Wolfgang Schuler and other masters of arts for their opinions in reference to the appearance of "a prodigy in the form of a star."[60] Observation had situated the prodigy above the moon and some men suggested that it foretold a golden age.[61] In response, Schuler elucidated pub-

56 Brahe, "Astronomiae instauratae progymnasmata," 3: 5–299.

57 Kathe, *Die Wittenberger Philosophische Fakultät*, 464.

58 Ibid., 3: 120–123.

59 Ibid., 3: 132–135.

60 Ibid., 3: 133.

61 Ibid., 3: 135. At the end of *Progymnasmata*, Tycho wrote that certain men linked the emergence of the new star with the prophetic predictions about the advent of a peaceful and harmonious age on earth before the universal conflagration. He himself conjectured that, even though there would be such an era announcing the destruction of all visible things,

licly that the phenomenon had given rise to a dilemma between trusting one's own observations and Peripatetic doctrine; he eventually resolved this difficulty by defining the nova as an unusual supralunary comet. By making the new star the result of an unusual divine intervention into the normally established order, Schuler argued that he could persist in Aristotelian teaching regarding the incorruptibility of the celestial realm: his observations legitimately contradicted Peripatetic doctrine since the comet was a divine exception to the ordinary divinely established rule of nature. The invocation of God's supernatural power was one approach taken by the sixteenth- and seventeenth-century Scholastic philosophers in order to preserve Aristotle's teaching on the incorruptibility of the celestial realm.[62] Schuler substantiated his interpretation by adding that the correctly established hierarchy of arts was not harmed in this way and that Holy Writ attested that such divine interventions had happened before, referring to the standstill of the sun in Joshua (Josh.10:12). The wonderful sign showed humankind that God was not to be bound by secondary causes.[63]

Tycho Brahe was harsh in his subsequent comments. He ascribed the opinions of the Wittenberg astronomers to their poor choice of instruments, their adherence to Peripatetic philosophy, and a biased reading of Scripture. On the subject of the instrument, Tycho explained that the astronomers at Wittenberg did establish a parallax (and consequently did not situate the new star in the upper region of the fixed stars) due to their use of a deficient wooden quadrant. When visiting Wittenberg, he expressed his surprise that, despite the high reputation of the university in the field of mathematics, he found that little care was taken in the development of new instruments and he saw the astronomers employing a wooden quadrant that did not yield exact results.[64] According to Tycho, Peripatetic inquiry of heaven implied turning away from the Book of Nature. University astronomers read Aristotle, not the Book of Nature.

> If only, on the particular occasion of this star, those who frequent the Academies, be it teachers or students, would for once make an effort to open their eyes and perceive from the Book of Nature itself with how

---

further knowledge about the timing and the ways in which this era would unfold was only given to God. For a reading of Tycho's own exposition, see ibid., 3: 312.

62 Grant, *Planets*, 212–213.

63 For Tycho's account of Schuler's exposition, see: Brahe, "Astronomiae instauratae progymnasmata," 3: 135–139.

64 Brahe, "Astronomiae instauratae progymnasmata," 3: 150.

> many great errors that heathenish Aristotle has bewitched them for so many centuries: [then] they would understand indeed that many things which they proclaim as certain and indubitable, should be held very differently. But I do not know by what fate of errors and darknesses clouding men it comes about that what is deviant is chosen and approved as correct, and what is plausible as appropriate, or, still further, trumpeted as authentic. In such a manner, this Aristotelian opinion on the generation and position of comets, despite being not at all consonant with experience, is still approved in all the academies, and bitterly defended to establish it. For they do not find it unpleasant or embarrassing to keep singing an old song or rather jingle, and reheat cabbage reheated so many times. Indeed, they consider that it is inappropriate to depart even a hair's breadth from the Aristotelian decrees, or ever to cast doubt upon them. O servitude.[65]

The intellectual servility of the Wittenberg astronomers was telling, in fact, of a larger critique of university education as a whole. The Book of Nature arose as an autonomous legitimate cognitive category next to authoritative books. On two other occasions in his comment on the Wittenberg account of the nova, the Lord of Uraniborg argued that these mathematicians had the habit of turning to books instead of heaven when practicing astronomy.[66]

By reading the Book of Nature through a pagan lens, the academies were prevented from understanding the divine signs properly and accessing ontological and divine truth. This truth was the imperturbable light of God. Truth, in Tycho's discourse, was the queen of all virtues:[67]

65 Ibid., 3: 141. "Utinam vero academiarum frequentatores sive docentes, sive discentes, vel huius Stellae occasione, oculos semel aperire, et ex ipso naturae libro, quot quantisque erroribus, tot iam seculis eos fascinaverit ethnicus ille Aristoteles, dispicere satagerent: equidem plurima, quae pro ratis et indubitatis quotidie personant, multo aliter se habere intelligent. Sed nescio, quo errorum et tenebrositatum terricolas obnubilantium fato, id eveniat, ut devium pro recto, atque plausibile pro congruo, ut plurimum eligatur atque approbetur: imo hincinde tanquam authenticum ebuccinetur: quemadmodum etiamnum in omnibus academiis haec ipsa quoque de cometarum generatione et situ Aristotelica sententia, ut ipsi experientiae nequaquam consona, approbatur, proque ea stabilienda, acriter certatur. Neque enim cantilenam veterem aut potius naeniam decantare, crambemque toties coctam recoquere, illis ingratum, aut fastidiosum est. Imo nefas esse ducunt, ab Aristoteleis decretis vel latum unguem decedere, aut ea semel in dubium vocare. O servitus."

66 Ibid., 3: 140, 149.

67 Ibid., 2: 159.

> On the other hand, we deem that nothing is to be brought against those who have tried to criticize and weaken the very observations shown here as principles, and the geometrical demonstrations susceptible to no hesitation, either out of ignorance or [because of] the given data, or out of a malevolence and disdain of truth, and a love of ancient errors. Those who do not want to know what the matter truly is, let them rejoice in emptiness and errors and be blind in the darkness, and let them please themselves who neither want to nor are able to behold the light of God. The truth remains unmoved by itself, and triumphs in the end, whether human beings pursue and approve it or not.[68]

The Book of Nature—displaying the visible effects of divine providence—became clouded in shadows when privileging a human pagan author over the authority of the divine. One should turn to God instead of to a misconceived demi-god. The truth stood independent from the human mind, regardless of its assent or disapproval. This contrast between the mind and the truth reflects the contrast between uncertain authoritative opinion and everlasting mathematical certainty. In Tycho's view, the "naked truth" was intact and incorruptible.[69] He frequently used the imagery of darkness and light, blindness and seeing, to depict the relation between truth and opinion. Truth was not a construction of the human mind that was wrapped in shadows. Hence, Tycho's claim that his methodology enabled God's light to shine. What was truly at stake in Tycho's wrestling arena was the true advocacy of the divinely authored Book of Nature.

3 The Theological Anthropological Turn

Tycho's account of the 1572 star echoed a clear theological anthropologically founded plea for a pious astronomy. His opposition against the Peripatetics was an obvious example of the increasing tendency in the late Renaissance

68 Ibid., 3: 382: "Nos autem contra eos, qui observationes ipsas tanquam principia hic adhibitas, et geometricas demonstrationes nulli haesitationi obnoxias, sive per inscitiam, sive data opera, aut malevolentia quadam atque veritatis odio, errorumque veterum amore, carpere atque elevare attentarint, nihil disserendum censemus. Gaudeant vanitatibus atque erroribus, qui id quod res est, cognoscere nolint, et in tenebris caecutiant, illicque sibi placeant, qui lumen Dei intueri, nec volunt nec possunt. Veritas per se immota manet, sive homines eam assequantur approbentque, sive non, tandemque triumphat."

69 Ibid., 2: 5.

to articulate the nature of a pious philosophy by opposing it to specific representations of the adherents to Aristotelian natural philosophy.[70] The piety discourse became the backbone of Tycho's representation of the Peripatetic philosopher.

3.1 *Observing Divine Providence*

In Tycho's writings, celestial observation merged scientific observation and religious observance. In the 1574 Copenhagen oration, Tycho clearly attributed religious meaningfulness to his stargazing activity: the learned astronomer and God were involved in a continuous relation, a relation which could be actualized, reinforced or weakened through the study of divine creation. Gazing upon the stars represented this symbiosis of observation and observance, just as the new star had been the object of both observation and observance. The astronomer was an observer of general and special divine providence.

If one was to notice anything new in the skies, continuous observation was key, according to the Danish astronomer. By his own account, he himself had been contemplating the heavenly theater since childhood and possessed a great familiarity with it.[71] Failing to study the stars by means of immediate observation implied the risk of missing out on divine miracles in heaven, like the Star of Betlehem[72] or the new star that appeared in the constellation of Cassiopeia in 1572. The *nova stella* was a "very rare miracle" because "the very long experience of many centuries has taught that Heaven with all its bodies remains unexposed to any change since the first Creation."[73] He thought of the nova as an instance of special providence, meaning an unusual divine intervention contrary to the ordinary course of nature (general providence): God had chosen to transcend the usual laws of nature.[74] Some like Caspar Peucer insisted that this miracle announced the end of the world and the second coming of Christ on earth, foretelling the "glorious liberation of the awfully distressed church."[75] Over time, Tycho's opinion on this matter seems to have

---

70 Blair, "Mosaic Physics," 34.

71 Brahe, "Astronomiae instauratae progymnasmata," 3: 97.

72 The three wise men from the East noticed the Star before other mortals, precisely because these wise men devoted themselves more abundantly to the contemplation of the stars. The star would eventually lead the Magi to Jesus in Bethlehem, and thus announced the first coming of Christ on earth; see: ibid., 2: 337.

73 Ibid., 3: 310; 2: 320: "Coelum enim cum omnibus suis Corporibus nulli mutationi obnoxium inde a prima Creatione permanere, diutissima multorum Seculorum docuit Experientia."

74 On the argument of special providence in the sixteenth-century interpretation of the nova, see: Methuen, "This Comet," 499–515; eadem, "Special Providence," 99–113.

75 Brahe, "Astronomiae instauratae progymnasmata," 3: 121: "[...] gloriosam liberationem

changed. In 1573, he apparently agreed that the new star was the herald of the end of time, and considered this interpretation to be in line with God's freedom and power to do as He pleased:

> It is left for us to state that we should marvel at this prodigy of God, the artisan of the entire world-machine; He placed it at the beginning beyond the entire order of nature, and finally it is shown to the world as it comes to an end. For the divine majesty acts with the greatest freedom, and is not confined by any chain of nature, and if He wants He stops the water in rivers and turns the stars backwards.[76]

By the 1580s and early 1590s, when the *Progymnasmata* was written, his mind seems to have changed, and he now argued that knowledge about timings and the relevant signs was a prescience given only to God.[77]

The astronomer's role was to be a trustworthy reporter of divine providence. On several occasions, Tycho indicated that he wanted to provide posterity with a certain account of the 1572 miracle.[78] In order to assess the exceptionality of the miracle, he had read Biblical and non-Biblical reports of the new star. A comparison with other miracles noted down in the Bible had shown that the nova was a very rare miracle of the quality of the solar eclipse during Christ's crucifixion but not so unique as the Bethlehem star witnessed by the Magi. By contrast, Tycho displayed a highly critical attitude towards the secular reports on extraordinary celestial phenomena, finding them either insufficiently substantiated or affected by prejudice. In his opinion, the biased interpretation of celestial phenomena due to an adherence to Peripatetic philosophy had affected previous historiographical writing on the special effects of divine providence. Only one secular (i.e., non-Scriptural) historical record attested to the coming-into-existence of another new star, and the reception of this record by posterity had become tainted by prejudice. Tycho was referring to the sixth chapter of the second book of Pliny the Elder's *Naturalis Historia*, in which he mentioned Hipparchus's observation of what the Greek astronomer believed to

misere afflictae Ecclesiae." Into his account of the new star, Tycho copied a letter from Peucer to Wilhelm IV, dated 1 January 1573.

76 Ibid., 3: 99: "Reliquum igitur est, ut statuamus Dei totius machinae mundanae opificis, admirandum hoc esse ostentum, praeter omnem naturae ordinem, a seipso in initio constitutum: nunc demum advesperascenti mundo exhibitum. Divina enim maiestas liberrime agit, nec ullis obstricta est naturae vinculis, sed cum vult, sistit aquam fluviis et vertit sidera retro."

77 Ibid., 3: 313.

78 Ibid., 3: 300.

be a new star. No similar events had been reported since Hipparchus's observations.[79] Tycho held Hipparchus in high esteem; Hipparchus's portrait hung in Uraniborg castle as the first of a series of portraits of great astronomers throughout history.[80] Tycho said that he knew of interpreters who had distorted Pliny's words, stating instead that Hipparchus had in fact seen a comet. This misrepresentation was due to the interpreters' desire to stay close to the Aristotelian doctrine in which they had been educated since childhood.[81] None of these translators was identified by Tycho apart from a certain Milich. This must have been Jakob Milich (1501–1559), professor at Wittenberg University in the first half of the sixteenth century who had written a commentary on the second book of the *Historia Naturalis*.[82] The distortion of Pliny's words produced a Scholastic bias in the retelling of Hipparchus's account.[83] Even though Tycho himself considered it very probable that Hipparchus was capable of discerning a new star, he could not fully accept the truth of Pliny's account in the absence of other reports.[84] He hinted however that his own investigation into the nature of the phenomenon had gone some way to support the trustworthiness of Pliny's account. After some joint observations conducted in Copenhagen, observational evidence had convinced the Danish professor of medicine Johannes Pratensis (1543–1576) that his fellow observer's claim to have seen a new star was serious and he came to realize that Pliny's account of Hipparchus observing a new star was not false.[85] Tycho represented himself (and the mathematical astronomers generally) as a trustworthy reporter of divine providence. The absence of historical records on similar stars in the past, apart

---

79 Ibid., 2: 323.

80 Blair, "Tycho Brahe's Critique," 356. The series consisted of portraits of Ptolemaeus, Timocharis (c. 320–260 BCE), al-Bitruji (fl. 1200), the Spanish King Alphonse X (1221–1284), Copernicus, Tycho himself, and his son Tychonides.

81 Brahe, "Astronomiae instauratae progymnasmata," 2: 322.

82 Ibid., 3: 98; Thorndike, *History of Magic*, 387–388. Milich's commentary was part of a "Protestant" line of commentaries on Pliny's *Naturalis Historia* as part of his instruction at Wittenberg University; see: Nauert, *Humanism*, 80.

83 Brahe, "Astronomiae instauratae progymnasmata," 2: 322; Ibid., 3: 98. Tycho also adduced an observational argument in favor of Pliny's account of Hipparchus's observation. Milich adduced that Hipparchus would have determined the position and motion of the fixed stars by means of this comet. In Tycho's opinion, this is ridiculous, for the enormous, unstable, fast and unsteady course of the comets stands in stark contrast to the regular, steady, uniform, and extremely slow motions of the fixed stars in the eighth sphere.

84 Ibid., 2: 231–322.

85 Ibid., 3: 94.

from Pliny's account of Hipparchus, did not necessarily indicate that there had been no other new stars. This could also be the outcome of the historiographer's theoretical biases (leading him to mislabel the type of celestial event observed), and the concomitant *a priori* judgments on the regularity, irregularity, and even possibility or impossibility of such an event. As a result, similar incredible or wondrous phenomena might have passed unnoticed or were misrepresented.[86] Tycho's lack of confidence in earlier historiography was also evident from his dealings with four secular historical reports on transient stars.[87] He questioned the dependability of the accounts for various reasons, ranging from proper identification of the celestial phenomenon, and the lack of historiographical unanimity on the occurrence of events, through to the anonymity of the respective sources of information. At the end of his critical evaluation, Tycho ruefully remarked as follows: "And that is why it is to be wished that the appearances of this [star] were chronicled more fully by a suitable astronomer."[88] This desideratum is precisely what defines Tycho's program for *Progymnasmata* covering the miracle of 1572. The observational mathematical astronomer now emerges as an excellent historiographer of heaven for he proceeded in an unbiased and accurate fashion.

The third chapter of the *Progymnasmata*, introducing the theme of the new star, states that close observation was the only way to proceed, and that this should be undertaken cautiously and, most importantly, piously.[89] That this direct and continuous observation was absent at universities was emphasized through Tycho's turn to the antitype of the learned observer—what I would call the naïve or unspoiled observer. Tycho recounted that unlearned men had been known to spot new stars prior to the educated astronomers, despite their lack of observational practice. When confronted with the extraordinary events in 1572 and 1577, Tycho himself sought confirmation from servants and peasants

86 Ibid., 2: 329.

87 Ibid., 2: 327–328. The respective records concern Flavius Josephus's account of a transient star which, shaped like a sword, stood visible above Jerusalem for a year and announced the destruction of the city; secondly, the statements of "certain historians" regarding the appearance of a new star during the reign of Emperor Hadrian (76–138), and, finally, two accounts by Cyprianus Leovitius (1524–1574) on a star in 1264. Tycho questions the accuracy of these accounts based on the lack of historiographical unanimity, the anonymity of the respective sources of information, and the inability to properly identify the observed phenomena.

88 Ibid., 2: 329: "Atque optandum foret eius apparentias per idoneum astronomum fuisse uberius literis consignatas."

89 Ibid., 2: 311–312.

who had but a mediocre knowledge of the stars.[90] His use of these witnesses as supports for his own observation confirms Richard Serjeantson's opinion that the early modern dismissal of human authority did not apply to human testimony (that was considered to be a form of authority).[91] Tycho's reporting on these men as witnesses and their testimonies served a rhetorical, persuasive purpose.[92] For the miracle of 1572, these witnesses were sailormen, couriers, coachmen, and similar—men who were "illiterate, uncultured and not trained in the Academies."[93] The year-round experience of these men with the visible surface of the heavens primed their authority as witnesses: the *illiterati* were those men who were dependent on observation of the heavens for professional reasons, but lacked academic training:

> Because cart-drivers, travelers and seamen spend most of their time under the open sky, and travel abroad just as often by night as by day, it is not surprising that they realize that there is something new in Heaven, more quickly than those who hide in heated rooms as much as possible, and, like snails, rarely emerge from their shells, and when they do, look at the earthly things in front of their feet, just like moles, scarcely raising their faces to that extraordinary sky, and giving much less attention to the heavens than they deserve.[94]

Tycho's naïve observer argument reinforces the importance of an immediate engagement with heaven even if they could achieve no more than a primitive knowledge of the new star.[95] He criticized the astronomers who no longer looked up to stars as displaying an attitude contemptuous of God's gifts.[96] In his opinion, there was something miraculous in the fact that the illiterate turned

90 Ibid., 2: 308.

91 Serjeantson, "Testimony," 197.

92 Westman, *Copernican Question*, 233. The role of these naïve witnesses was considerably less elaborate in the 1573 account of the new star.

93 Brahe, "Astronomiae instauratae progymnasmata," 2: 311.

94 Ibid., 2: 312: "Cum enim aurigae, viatores et nautae, plerunque sub dio versentur, atque peregrinentur, idque saepenumero tam noctu, quam inter lucano tempore, non adeo mirum esse debet, eos citius aliquid noui in coelo persentire, quam hos, qui in hypocaustis ut plurimum latitantes, instar cochlearum e suis testis vix egrediuntur, aut si id fiat, terrestria potius, et quae ante pedes sunt, more talparum solummodo intuentur, in supereminentem illum zethera vix et ne vix quidem vultus attollentes, multo minus ut debita consideratione coelestia scrutentur."

95 Ibid., 2: 311.

96 Ibid., 2: 316.

out to see more than the majority of the astronomers.[97] He elaborated this theme through the myth of Cassiopeia, after whom the constellation in which the new star had become apparent was named. According to Tycho's account, Cassiopeia was an Ethiopian queen, renowned for her extraordinary beauty, who had the audacity to despise the beauty of the goddess Juno and the water nymphs the Nereids. Due to her display of excessive arrogance, Juno warned Cassiopeia that irrespective of finery or excellence of appearance that mortals might have, these things are to be disdained compared to those that are divine. Cassiopeia persisted in her conceit, however, and, as a subsequent punishment of the gods, her realm was overwhelmed by disaster.[98] According to Tycho, the myth taught that astronomers must refrain from arrogance and pride before the gifts of God and "haughty disdain" for God Himself. Their behavior had to display a simple and inward piety.[99] In the light of his exposition of the myth and the remarks about the miraculous nature of the ignorant being the first observers of the star, Tycho's reproach of impiety seems to suggest that the new star in the constellation of Cassiopeia was a divine warning to those blinded by earthly desires that they should refrain from insouciance and pride.[100] Not disdaining divine gifts here means looking up to heaven and acquiring first-hand knowledge of the stars, engaging in an immediate experience of divine creation. Moreover, the learned forsook their responsibility to keep the ignorant from straying into darkness.[101] Tycho argued that it was not surprising that the simple and illiterate mass, in thrall to the learned who disdained divine gifts, should ignore and neglect the path of truth and virtue. The eternal and true were thereby neglected. This path of truth and virtue, Tycho affirmed, is the hardest to follow "for the corrupted nature of the earthly Adam."[102] The responsibility of the learned towards the community reminds us of Rheticus's claim that the true church had to be educated in order for its members not to stray from the true path. He was equally convinced that the learned should light the way for the unlearned.

3.2 *Confessing Ignorance*

The possibility that, by their intellectual behavior, scholars could be committing an intellectual vice—failing to acknowledge the possibly deceitful

97 Ibid., 2: 311.
98 Ibid., 2: 312–313.
99 Ibid., 2: 311–312.
100 Ibid., 2: 316.
101 Ibid.
102 Ibid. "[…] natura Adami terrestris corruptae."

nature of one's reason—, already incipient in Rheticus's *Epistola de terrae motu*, became ever more palpable and targeted at Scholastic natural philosophy in Tycho's self-published, extra-academic writings. He was not caught up merely in a "secular" or "social" difference of opinion between different philosophical schools. The Danish lord was instead portraying the Peripatetic as acting impiously before God when reading His divinely authored book.

"Indisputably, so great is the majesty of the best and thrice great God, that the wisdom of His works themselves fails to be exhausted by any creature," wrote Tycho in the *Mechanica*, in his dedication to Emperor Rudolf II.[103] A pious attitude to the study of the divine creation was built on the presumption of the unpredictability of God's freedom, as seen in the miracle of 1572, and by admitting the limited knowability of divine omnipotence when studying the regular effects of God's providence. Like Rheticus, Tycho shared Luther's notion of metaphysical ignorance: true knowledge was reserved for the visible realm, ignorance prevailed in the invisible realm. This epistemological pessimism seems to be at variance with Tycho's opposition to the Scholastics. At first glance, his proceedings attest to a generally optimistic appraisal of the abilities of human reason, based on certitude. Still, the Danish astronomer ultimately tended to refrain from making positive claims about the celestial realm. His inquiry into these extraordinary phenomena had proven that the region above the moon was not incorruptible but was susceptible to change. Likewise, he established that the observed trajectory of the comets through the ethereal realm implied that there were no solid celestial spheres making up the celestial realm, but he did not postulate what was the celestial matter that did cause the planetary motions. The historiographical understanding of Tycho's disciplinary incursions has already been refined in this respect. Kenneth Howell states that, in his study of heaven, observations could prove hypotheses wrong [i.e., could falsify them] but could not positively establish them.[104] Steven Vanden Broecke, moreover, points out the "consistent negativism" of Tycho's mathematical operations. The astronomer lacked the ability to generate mathematical alternatives for his disruption of natural philosophical accounts.[105] Hence, his theoretical activities could do no more than rebut existing claims. That a true role reversal did not ensue from Tycho's disciplinary crossovers was here explained from a theological anthropological perspective. His skeptical remarks show that maintaining the continuous relation with God in a pious

---

103 Brahe, "Astronomiae instauratae mechanica," 5: 5: "Tanta nimirum est Dei optimi et ter maximi majestas, ut a nullis creaturis ipsius operum sapientia exhauriri queat."

104 Howell, *God's Two Books*, 76, 95.

105 Vanden Broecke, "Teratology," 9.

way implied admitting one's epistemic limitations in the face of divine omnipotence. Two characteristics stand out in his skeptical discourse, the first being that ignorance is a feature with respect to the invisible part of nature (the invisible cannot be quantified, and quantification—or revelation—is a precondition of certainty), and the second is that references to the absence of knowledge were made with regard to divine omnipotence (or wisdom) and, as such, point to a human condition.

Experience in Scholastic natural philosophy was seen as a general statement about how the world behaved. Such statements were rooted in common opinion, the positions of philosophers, and sense-perception.[106] By contrast, Tycho advocated the importance of immediate sense-perception as a source of certain knowledge about the world, and hence of credibility. His mathematical analysis of the new star concerned its position, distance, and size.[107] On the other hand, the uncertainty that Tycho found characteristic of the invisible or barely visible aspects of celestial phenomena, and thus not susceptible to mathematical demonstration, is manifest in several places. On the subject of the coming-into-existence of the nova, for instance, Tycho wrote in 1573 that they "are not exposed to the senses and to Geometrical Demonstration and have principles that deviate from the pure certitude of the Mathematical Disciplines, [that are] even little verisimilar or outright conjectural."[108] This ignorance applied to the mathematician, to the natural philosopher, and to the theologian: it was a human condition of ignorance before divine omnipotence:

> Let all philosophers, both old and new, be silent: let even the theological interpreters of the divine mysteries be silent: let the mathematical contemplators of the celestial bodies be silent, nor think themselves able to state anything certain on the mode of generation of this star and the reason for such a miracle [...]. For it suffices to have shown that this new and unseen star, which has recently appeared, has no connection with that which was observed by the wise men; nor can its mode of generation be saved, either by the theologians, or by the philosophers, or even by the mathematicians themselves.[109]

---

106 Dear, "Totius," 148.

107 Brahe, "Astronomiae instauratae progymnasmata," 2: 330, 3: 112.

108 Ibid., 3: 304: "[...] siquidem sensibus et demonstrationi geometricae non subjaceant, et ab intemeratz mathematicorum certitudine, dissentanea habeant principia, imo saltem verisimilia aut plane stochastica [...]."

109 Ibid., 3: 99: "Taceant igitur omnes philosophi, seu veteres, seu novi: taceant ipsi quoque divinorum mysteriorum interpretes theologi: taceant coelestium corporum contemplatores mathematici, nec de modo generationis huius stellae et tanti miraculi excusatione,

An excellent metaphor, supplied by Tycho himself in reference to the Peripatetic philosophers, to describe this human condition of ignorance is the abyss of nature. Those men who were "soaked" in Aristotelian doctrine (and who occupied almost all chairs in the academies) never had any doubts about Aristotle's opinion; whenever they heard something contrary to Aristotle's opinion, they would subject these departures from Scholastic norms to ridicule and rejection. Their adherence to authority, Tycho explained, prevented them from finding something more true, and this he described as an outcome arising from the fallibility of the mind: "The abyss of nature after all is inexhaustible and the human mind very often falters and is clouded in matters distant from the external senses."[110] The notion of the abyss points to the unfathomable nature of nature itself: the readability of the Book of Nature, the comprehensibility of divine wisdom, only went so far as the senses allowed.

As a recognition of one's human condition, one had to display a humble attitude before divine creation. In his German treatise on the comet of 1577, Tycho's argument on the unknowability of the matter, and on the generation and justification of an extraordinary phenomenon such as a comet is turned into a larger statement about the deficiency of human understanding of the regular invisible causes. From this perspective he asked the philosopher to humble himself before divine providence and admit his ignorance. Humility, to quote the French contemporary philosopher André Comte-Sponville, emerges here as "la vertu de l'homme qui sait n'être pas Dieu."

> Whether they [the comets] are ordained by God and fabricated by Superior Penates unknown to us, or whether God the Almighty through His power and will and in His own good time makes such a new light in the heavens without [material] means, in order to warn us of future punishment, is not to be disputed at length here, for we humans, with our limited and earthly understanding, really have no proper grounds or conceptual framework for explaining the material of comets and how they are generated in a way which would not seem a wonder to us. We have no real knowledge of the matter or nature of the whole heavens, sun, and moon, nor what causes their wonderfully adroit motions, though they

se aliquid certi constituere posse, existiment [...]. Sufficit enim demonstrasse hanc novam et inusitatam stellam, quae nuper apparuit, nullam habere cognationem cum illa, quae magis conspiciebatur: nec posse eius generationis modum saluari, vel a theologis, vel a philosophis, nec ab ipsis etiam mathematicis."

110 Ibid., 2: 373: "Inexhausta enim est naturae abyssus, et mens humana in plurimis praesertim adeo longe a sensibus externis sepositis, saepenumero caligat atque titubat."

> have stood and been visible since the beginning of the world. How many things there are, even here on earth, which we see with our eyes and seize with our hands, the nature of which we can never satisfactorily comprehend. Therefore, philosophers should not contend so uselessly about things they cannot resolve, but should the more readily admit to a modest ignorance and say that comets are a special creation of God which come from unknown natural causes, of which we do not know how they are born.[111]

The Swiss physician and occultist Paracelsus (1493–1541) had argued that new stars and comets were created in heaven by Superior Penates—intermediate powers between God and man—when they were ordained to do so by God. Given his affinity with Paracelsian philosophy, it would not have been entirely surprising if Tycho had aligned himself with Paraselsus.[112] Rather than so doing, however, Tycho urged both Aristotelian and Paracelsian philosophers to admit their ignorance before God's omnipotence and to state the unknowability of the generation of the comets as well as of celestial matter and the causes of the planetary motions. According to Adam Mosley, Tycho displayed a conservative attitude here with respect to the power of mathematics to inform cosmology.[113]

Tycho reaffirmed his opinion on the unknowability of celestial matter in his correspondence with Christoph Rothmann, court mathematician to Landgrave William IV at Hessen-Kassel. In response to Rothmann's claim, based

---

111 Christianson, "Tycho Brahe's German Treatise," 133; Brahe, "Vonn der Cometten," 4: 383: "[...] dann ob schon aus gottes verhengcknus die penates superi, die vnns vnbekandt, solches fabricieren, oder obs auch gott der Allmechtig ohne mittel durch seine krafft vnnd willen zu seiner zeit ein sollich neu liecht am himel vnns zur warnung zukonfftiger straff schaffet, ist nicht nott alhier weitleufig dauon zu disputieren, die weil wir menschen durch vnnsern ringen vnnd irdischen verstandt eigentlich kain rechten grundt vnnd wissenschafft haben konnen, was die matteria Comettarum sei, vnnd wie si genneriert werde, dass vnss auch nit wunder nemen soll, die weil wir des ganntzen himmelss Son vnnd Mon, die doch allzeit vom anfang der welt gestanden vnnd geschinen, was ir matteria vnnd wesen sei, ainiche wissenschafft, vnnd was inen en wunderbarlichen behenden lauff mache, nicht haben, ja wieuil ding sein hierunden auf dem erdtboden, die wir mit vnnsern augen ansehen vnnd henden greiffen, dessen nattur wir doch nimmer mehr genuegsamb konnen lernen erkennen. Darumb sollen die philosophi nicht so vnnützlich streitten von den sachen, die si nit zuermessen wissen, sonndern vil mehr vnnsre ignorantia modeste bekennen vnnd sagen, das die Cometten seien ein sonnderlich geschepff gottes, das auss verborgenen vrsachen der nattur kombt, welches vnns vnbekanndt ist, wie es geboren wirt."

112 Howell, "Role of Biblical Interpretation," 515.

113 Mosley, *Bearing the Heavens*, 72.

upon his research into refraction, that the entire universe is filled with air, Tycho stated that the heavenly substance is unknown. All he knew was that heaven consists of a matter that was "very fluid, very thin and very delicate that allowed the free motions of the seven planets."[114] In arguing as such, Tycho formulated the inability of mathematical astronomy to generate a positive stance on this part of physical reality. This negativism was more than a consequence of the astronomer's disciplinary role; it was extended to all mortal human beings as the result of the unfathomable wisdom of God.

According to Tycho, the discussion regarding celestial matter was one for the theologians and philosophers but they had not reached a satisfactory conclusion, owing to the feebleness of the human mind. In the end, celestial matter proved to be unknowable due to the human condition:

> Therefore, the question of the matter of heaven itself is not a decision that properly pertains to the astronomer; after all he does not strive to investigate what heaven is and what its splendid bodies consist of, but rather how all these [bodies] are moved, based on certain observations. The other matter is left for theologians and natural philosophers to discuss, among whom moreover no satisfactory explanation persists so far. The weakness of the human mind is so great, that it cannot entirely fathom God's creation, not even terrestrial things exposed to the external senses and so close that we can touch them.[115]

The scope of the astronomer's intellectual potential for certain knowledge was thus restricted. Tycho attributed the pertinent ignorance regarding the material cause of heaven to the asymmetrical relation between the deficient human intellect and God's unfathomable works. By arguing that the human mind could not even fathom the terrestrial works of God that were nevertheless unhidden from the external senses, the Danish astronomer shifted the question of celestial matter to the unintelligible aspects of the realm of terrestrial

114 Tycho Brahe, letter to Christoph Rothmann, Uraniborg, 17 August 1588, 6: 140: "[…] liquidissima, rarissima, et subtilissima substantia constare, quae septem Planetarum cursus liberos […] admittat."

115 Ibid., 6: 139–140: "Quemadmodum ipsa etiam de materia caeli quaestio, non proprie ad astronomi decisionem pertinet; Is enim non quid sit caelum, et unde ipsius splendida corpora constent, sed saltem quomodo omnia haec moveantur, ex obseruationibus certis pervestigare laborat. Alterum enim theologis et physicis discutiendum relinquit, apud quos etiam hactenus idipsum non satis explanatum permansit. Tanta enim est humani ingenii imbecillitas, ut opera universa Creatoris, ne quidem terrestria et sensibus externis obvia, nobisque adeo ut palpentur propinqua, penitus haurire queat."

nature. In spite of its unknown identity, Tycho did believe that celestial matter differed from elementary substance.

> But I believe that no man is allowed to set forth something certain about the matter of Heaven (If indeed it can be properly said to have a matter). For heaven is something abstract and similar to the immaterial, escaping our understanding; nonetheless, we may conclude that [heaven] itself does not participate in elementary nature. Because this [heavenly matter] is by far superior and more divine, and therefore [it is] something incorruptible and elevated above every elementary quality, and of a very pure, simple and transparent, and very rare substance, molded by the marvelous wisdom of God and inscrutable to us.[116]

Due to this *a priori* belief in a celestial nature that was distinctly different from its elementary counterpart, Adam Mosley does not deem the Danish astronomer "a true advocate of the skeptical position that physical knowledge of the heavens' true nature was unobtainable by man."[117] This is indeed one of Tycho's *a priori* convictions. Nonetheless, his distinction between the respective substances of the celestial and terrestrial realm clearly did not entirely suspend a skeptical position on the knowability of celestial matter.

Tycho's general program consisted in a humbling of the figure of the Scholastic natural philosopher, a figure in the grip of explanatory hubris and authority, for not recognizing his "earthly understanding." Despite the fact that man gathered knowledge about the celestial bodies ever since the beginning of creation, he could explain neither the cause of their movements nor the nature of their bodies. This ignorance was also felt regarding many other things, even terrestrial ones, in spite of the information that was gathered by means of experience (eyes and hands). The unknowable thus concerned the invisible and intangible domains. When reflecting about the world, the mind was subject to limitations in terms of the intelligibility of the world. Humans could acquire true knowledge only of the accidental aspects of phenomena, of that which could be sensibly discerned. Ignorance about the invisible causes of the phenomena

---

116 Ibid., 6: 136: "Verum licet nulli mortalium de ipsissima caeli materia (si modo e materia aliqua constare proprie dici potest) certi quid proferre in promptu esse credam. Est enim caelum abstractum quid, et immateriali simile, nostrum captum effugiens, nihilominus satis concludere licet, ipsum de nulla elementari natura participare. Cum haec sit longe praestantius et divinius quiddam adeoque incorruptibile et supra omnem elementarem qualitatem exaltatum e purissima, simplici et pellucida rarissimaque substantia, admiranda Dei sapientia et nobis imperscrutabili modo plasmatum."

117 Mosley, *Bearing the Heavens*, 90.

persisted, in spite of centuries of observation, due to the asymmetrical relation between the human intellect and divine omnipotence. Consequently, those who were particularly concerned to acquire causal knowledge of ordinary phenomena that had been visible for centuries, *in casu* the natural philosophers, ought to confess their ignorance and not pretend to be capable of yielding a positive claim about the invisible.

3.3 *A Philosophy of the Book of Nature*

Tycho Brahe was in the process of developing an *a posteriori* study of the Book of Nature, a study of nature that was at liberty to question and rebut prevailing philosophical views based on universal inductions. The divinely authored Book of Nature emerges as an autonomous cognitive category; its inferred conclusions provided a lens through which Holy Scripture could be read. Tycho's appraisal of the status of the Bible in celestial inquiry is dealt with when commenting upon the views on the new star of the Spanish physician Francesco Vallés (1524–1592). Vallés's *De sacra philosophia*, published for the first time in Turin in 1587, attempted a study of nature by means of Biblical exegesis.[118] Based on two Biblical verses (Eccl.18:1 "He Who liveth in eternity, createth all at once," and the above-mentioned Gen.2:2), Vallés had argued that nothing new could have come into existence since creation, for God's works persisted eternally. The new star had been present ever since creation but had not been apparent due to the density of the celestial orbs. In his opinion, heaven consisted of a very firm and very dense substance and encompassed orbs that were firmer and harder than any type of metal. Due to the motion of the orbs, the star had eventually become visible.[119] Tycho first refuted Vallés's competence as an astronomer—he had localized the nova in the knee of Cassiopeia instead of the back of her throne—and took him to task as an adherent to the doctrines of Aristotle. Moreover, the Spanish physician gave a too-literal interpretation of Genesis 2:2. Caspar Peucer (Sr.) had similarly adduced this verse when pondering whether the new phenomenon could be a star: we cannot, he wrote, affirm lightly that new stars are being created in the ethereal region for this contradicts the Biblical verse that God rested on the seventh day.[120] According to Tycho, the nova proved that Moses's statements did not imply that miraculous events could no longer occur in heaven; God was not enjoying an eternal

118 Crowther, "Sacred Philosophy," 412.

119 For Vallés's line of reasoning, see: Brahe, "Astronomiae instauratae progymnasmata," 3: 87–88. Tycho equally addressed the matter in one of his two letters to Peucer; see: Tycho Brahe, letter to Caspar Peucer, Uraniborg, 13 September 1588, 7: 133–134.

120 Brahe, "Astronomiae instauratae progymnasmata," 3: 125.

Sabbath.[121] Divine omnipotence was not to be restricted by the limitations of human mind; rather, His providential effects had to be awaited. Like Rheticus, Tycho voiced the idea that men were guided by the celestial phenomena, implying an *a posteriori* position in the intellectual inquiry into nature. Vallés left no room for miraculous works. By arguing that the star was not new, moreover, he weakened the impact of the miracle.[122] Tycho added that if the Spaniard had been educated according to a truer philosophy, one in accordance with the Book of Nature, and if he had not been blinded by Peripatetic philosophy, he would have judged differently:

> If he [Vallés] had been educated according to a more true physics and according to the Book of Nature itself, not so much written as exposed to the senses and external treatment (which is pre-eminently analogous to those matters of physics cited from it in the Holy Bible), assuredly in many cases he would have brought forward more suitable and less doubtful things, which had not been so long sought for, and he would have confirmed everything more firmly and more clearly, not so much by the authority of the ancients as by inspection and experience of nature itself.[123]

By contrast, Tycho's expectant attitude towards divine omnipotence, not tying God's freedom to preconceived judgment, implies that his study of nature did not restrict divine omnipotence to a determinist stance. The presumed analogy between the things observed in the Book of Nature and the text in Holy Writ grounded his position that the texts in the Bible were to be read in agreement with his previously established conclusions based on experience. It was "irrefragable experience" by which one learned that Moses's words required a different interpretation.[124] In the absence of revelation, God's Word became readable in this respect only by means of a scientific study of that other divinely authored book. One implication of Tycho's turn to the Book of Nature is that, whereas Melanchthon, in his discussion of the four ancient philosophical schools, used Holy Scripture as a lamp in assessing the piety of

121 Ibid., 3: 125.

122 Ibid., 3: 89.

123 Ibid., 3: 92: "Si e veriore physica ipsoque naturae libro, non tam scripto, quam sensibus et tractationi externae obvio (qui apprime analogus est, iis quae physice ex hoc in Sacris Codicibus citantur) instructus fuisset, utique in plurimis aptiora, minusque dubia, nec adeo longe petita attulisset, neque tam veterum authoritate, quam praesentis ipsius naturae inspectione, experientiaque, rem omnem firmius et dilucidius comprobasset."

124 Ibid., 3: 125.

the Epicurean, Aristotelian, Academic, and Stoic schools, Tycho attributed this illuminating function to that other divinely authored book. The Book of Nature was a lamp, because undiluted experience brought to him by that book taught him that specific Aristotelian tenets needed to be rejected due to their incompatibility with unlimited divine freedom.

On a more general note, Tycho's argumentation against Vallés was but one illustration that he did not consider the Bible a source of exhaustive knowledge about the world. He argued that Biblical language was accommodated to the intellectual level of uneducated men. In a 1589 letter to Caspar Peucer Sr., in response to the latter's development of a Mosaic philosophy (taking the literal text as source of knowledge), Tycho replied that the language used in Scripture was adapted to uneducated men and hence, its information about nature was restricted to immediate experience:

> [...] Moses wrote down this history of creation only for the rude and simple people, and not for those educated in the sciences of mathematics and physics, and examined those things that meet the eyes of men, not the things that are invisible or hidden [...].[125]

With respect to the debate regarding the existence of solid celestial spheres, Tycho averred that the relevant Biblical passages were ambiguous on the subject.[126] However, this obscurity of Scripture could be lifted. Indecisiveness was not the final stop, as it was in Rheticus's hermeneutics of obscurity regarding the motion of the earth. Tycho confidently claimed that Moses's verse had to be read differently because of irrefutable experience. Hence, he modulated his reading of the Bible to his conclusions derived from observing nature. The appearance of a new star in Cassiopeia had taught him that Biblical references to the eternal preservation of divine works and the suspension of divine activity from the seventh day of creation onwards did not apply to miraculous works. Neither in nature nor in Scripture were divine freedom and omnipotence to be restricted to the mind.

And yet, compliance with Scripture did seem to matter to Tycho. In an earlier letter to Peucer, dated 13 September 1588, he described his great satisfaction when his correspondent supplied him with a reading of two Biblical passages

125 Tycho Brahe, letter to Caspar Peucer, Uraniborg [June 1589/1590], 7: 231: "Quare cum Moses hanc creationis historiam rudi et simplici tantum populo, non mathematicae aut physicae scientiae gnaris, conscripserit, eaque duntaxat quae in oculos hominum incurrunt, non quae invisibilia aut abscondita sunt [...]."

126 Blair, "Tycho Brahe's Critique," 363.

that sustained his position on the fluid substance of the heavens.[127] Peucer had provided him with an alternative explanation of a commonplace quote in the sixteenth-century debate about the celestial spheres, namely Job 37:18 ("Hast thou with him spread out the sky, which is strong, and as a molten looking glass?"). Tycho reported to Peucer that Vallés had opposed the idea of a very fluid and permeable heavenly substance by means of this very line. Now, Peucer had provided him with an alternative explanation of the reference to "firmness", namely that it indicated the firmness, density, and steadiness of the motions of all celestial bodies. The Danish astronomer concluded his hermeneutical discussion on the subject of celestial matter by stating that:

> Hence, when we state with certainty that the material of the heavens is very liquid and thinner and subtler than the whole air, nothing in the Holy Scriptures or in philosophy opposes us.[128]

Scripture served as a secondary means in Tycho's research about the heavens, and was employed to confirm previously established findings in natural philosophy.[129] This is emphasized by the fact that Tycho's hermeneutics was neither extensive nor consistent, and that he himself did not initiate hermeneutical discussions regarding a specific philosophical position, apart from on the topic of terrestrial mobility that will be discussed further.[130] The Danish astronomer applied a variety of exegetical techniques when including Scripture in his study of heaven.

Tycho did use the Bible as a historical source of ancient astronomy. Besides establishing the miraculous nature of the new star by means of the mathematical analysis of positions and motion, the Danish astronomer offered a comparison of Biblical miracles in order to assess the degree of exceptionality of the miracle. The only explicit motivation to sustain the comparability of these miracles was founded on their occurrence in heaven and their opposition to the usual order of nature. In Tycho's opinion, the nova was not to be equated with "vulgar comets" but was "at least comparable" with two other miracles which had occurred in Joshua's time: namely, the standstill of the sun—a commonplace in early modern astronomical literature—and the backward movement

127 Tycho Brahe to Caspar Peucer, Uraniborg, 13 September 1588, 7: 133.

128 Ibid., 7: 135: "Nihil igitur obstabit vel e sacris bibliis vel philosophorum decretis, quin coeli materiam liquidissimam et omni aëre tenuiorem subtilioremque certo statuamus."

129 Granada, "Tycho Brahe, Caspar Peucer," 569.

130 Ibid.; Mosley, *Bearing the Heavens*, 158, 159.

of the sundial made by Biblical King Ahaz.[131] Another "at least comparable miracle" was the solar eclipse during the crucifixion of Christ. Tycho insisted that the new star was not comparable to the star that appeared to the three wise men on the occasion of the birth of Christ. Despite the fact that the star of Bethlehem was also an immediate work of God and transcended all laws of nature, this particular miracle exceeded the others. The Bethlehem star was removed far from the ether and was not subject to the diurnal universal revolution of the stars. In the 1573 treatise, he explained that the star was localized in the elementary realm of air.[132]

Tycho's plea for an immediate turn to the Book of Nature to acquire knowledge of divine providence may readily be compared to Martin Luther's opposition to the Scholastic tendency to read the Bible through a Peripatetic lens. If the word "theology" in Luther's tenet 50 of the *Disputatio contra scholasticam theologiam* (1517), stating that "Aristotle is to theology what darkness is to light," were replaced by "the Book of Nature," this stance would represent an excellent summary of Tycho's opposition to the Peripatetic interpretation of the nova of 1572.[133] Based on their respective mistrust of the human mind, both Luther and Tycho perceived the writings of Aristotle as clouding a correct reading of the divinely authored books. Whereas the former resisted "a philosophical contamination of the Word," the latter opposed a Peripatetic determinist interpretation of heaven.[134] Both men were concerned with letting the light of God shine when reading His books. They wanted to establish a reading that avoided the pitfalls of the human mind. In Luther's case, this wish manifested itself in privileging the literal-historical sense when reading God's Word. Tycho, on the other hand, emphasized the primacy of mathematical astronomy in the study of heaven. Both men advocated their respective methodologies as the appropriate and pious way of reading a divinely authored work, without committing an intellectual vice. They propagated an individual reading, unhindered by authority and tradition. In very much the same way that Luther emancipated Scripture from human authority and tradition, Tycho's methodology allowed

---

131 Brahe, "Astronomiae instauratae progymnasmata," 2: 317. The story of the sundial may be summarized as follows: King Hezekiah, king of Judah and successor to King Ahaz who had a sundial made in his lifetime, was at death's door when the prophet Isaiah came to see him. The prophet told him to get his house in order for he would not live but die. Hezekiah wept bitterly and prayed to God. Isaiah told the king of Judah that his prayers had been heard and his life would be prolonged by at least fifteen years. As a divine sign, the shadow on his father's sundial moved ten degrees backwards; see: 2 Kings 20:8–11.

132 Brahe, "Astronomiae instauratae progymnasmata," 3: 99.

133 Büttgen, *Luther et la philosophie*, 60.

134 Ibid., 16.

the Book of Nature to be emancipated from the determinism of Peripatetic natural philosophy. For theological anthropological reasons, Tycho's disciplinary incursion resulted in either negative or probable claims but did not result in positive claims about the universe. This incapacity for positive claims implies that he still required an existing corpus of natural philosophy. Tycho limited the potential for certainty or true human knowledge to that which man could establish about the visible, about the evident which obtruded itself to human awareness. The Danish astronomer read the Book of Nature in very much the same way as Luther read the Bible: both men were carrying out a hermeneutical operation of stating the evident, that is, of giving a historical-literal interpretation of the divine text.

The existence of clear parallels between the early modern hermeneutical strategies of reading the respective books of Nature and Scripture has been established before.[135] For the Protestant world in particular, Peter Harrison argued that the Reformist approach to texts was a catalyst in the rise of early modern science because new approaches to the Bible might have prompted new approaches to nature.[136] Rather than arguing for the influence of Biblical hermeneutics on the study of nature, our present exposition points instead to the striking parallels between both hermeneutical fields. After all, the wish to clear away corrupt readings of texts, and to inquire into what was actually there, strongly takes after humanist methodology as well.[137] In this sense, Tycho can also be said to have performed a return *ad fontes*, this time with respect to the Book of Nature instead of literary manuscripts. I myself would rather make a case for the theological anthropological source of both Tycho's and Luther's preferred hermeneutical modes. The awareness of the limitations of the deceitful postlapsarian mind prompted the search for a method which decreased the possibility of deceit, and whose results were in agreement with these limitations.

4 The Geoheliocentric Model of the Universe

Given Tycho's admission of the incomprehensible nature of the invisible particulars of the divine works, and his mistrust of the human mind (manifested in his disparagement of the Peripatetic philosophers), one should wonder whether the Danish astronomer considered his geoheliocentric model of the

135 Crowther, "Sacred Philosophy," 400.
136 Harrison, *Bible*, 8, 10.
137 Methuen, "Interpreting the Books of Nature," 211, 213.

universe to be an ontological representation. Did this model correspond to reality, in Tycho's opinion? After all, astronomical hypotheses represented the invisible efficient causes of the planetary motions, about which Tycho explicitly denied the possibility of true knowledge in the first treatise on the nova. In light of the above discussion, it seems more plausible that he advanced a probable claim which approximated truth as far as possible. This suggestion solves some of the inconsistencies which are otherwise detected in his texts. As with Rheticus, Tycho's position would reflect a theological anthropological condition of temporal ignorance about the invisible causes of nature which would ultimately be lifted by revelation. This interpretation is in contradiction with the usual historiographical view according to which the Danish astronomer did think of his model as a physical reality, a view which has not been questioned so far, at least to my knowledge, even though I have not come across any unambiguous statements from the astronomer to this effect.[138]

The particulars of his geoheliocentric model are discussed in the famous eighth chapter of *De phaenomenis*.[139] Tycho localized the immobile earth at the center of the universe. The earth was orbited by the sun and the moon. The sun in turn was the center of the motions of the five remaining planets, Mars, Venus, Mercury, Jupiter, and Saturn. A geoheliocentric version of the planetary motions was developed already in the fifth century by Martianus Capella (*fl.c.*410 CE) who made Venus and Mercury orbit the sun. This Capellan system was included in an astronomical handbook printed in Venice in 1573, authored by Habsburg astronomer Valentin Naibod (1523–1593).[140] Given the fact that other sixteenth-century astronomers were experimenting with geoheliocentric models (or variations of the heliocentric model), we may doubt whether this system was indeed a Tychonic model.[141] It could very well be that around 1550 Erasmus Reinhold was already pondering the possibility of a geoheliocentric model, as some citations in his unpublished commentary on Copernicus's *De revolutionibus* (*Commentarius in opus revolutionium Copernici*) lead us to suspect, but there is as yet no further evidence.[142] Furthermore, there

138 For instance, see: Howell, *God's Two Books*, 76, 92; Schofield Jones, *Tychonic and Semi-Tychonic World Systems*, 311.

139 Brahe, "De mundi aetherei recentioribus phaenomenis," 4: 156–171.

140 Gingerich and Westman, "Wittich Connection," 47; Westman, "Three Responses," 321–322.

141 For an overview of other astronomers, see: Schofield Jones, *Tychonic and Semi-Tychonic World Systems*, 108–167.

142 Alexandre Birkenmajer claimed that Reinhold gave the first hint of a proto-geoheliocentric model in his commentary on *De revolutionibus* and that it was Reinhold, not Tycho, who was the first to assert the physical reality of the geoheliocentric model; see: Birkenmajer, "Le commentaire inédit," 175–177. In reply, Christine Schofield argued that the rele-

are distinct similarities between the work of Tycho and other contemporary astronomers such as Paul Wittich (c. 1546–1586) who visited the former for four months at Hven in 1580, and who, in his copy of *De revolutionibus*, had inserted a diagram entitled "From the Copernican hypotheses accommodated to the immobility of the earth." This copy contains Wittich's own annotations and diagrams as well as copies of Reinhold's annotations of *De revolutionibus*.[143] There was also a heated controversy between Tycho and the Habsburg astronomer Nicolas Reimers Baer or Ursus (1551–1600) whom the former accused of plagiarism.[144] Putting the issues of priority and exclusivity aside, the present section questions whether, given his previously mentioned indications of ignorance, the epistemological attitude of temporal causal ignorance argued for Rheticus could be applied to Tycho as well. Following an exploration of Tycho's appraisal of the relation between revelation and reason, the question will be addressed whether he himself gave hints of the hypothetical nature of his system.

4.1 *Revelation over Reason*

In the light of his opposition to the Peripatetics, Tycho advanced an inconsistent and problematic argument against Copernicus's position on the mobility of the earth.

> I hold that it should be stated beyond all doubt, following the ancient astronomers and the received opinions of the natural philosophers—and above all by the testimony of the Bible—that the earth on which we live occupies the center of the universe, and turns with no yearly motion (as Copernicus believed).[145]

In *Progymnasmata*, Tycho likewise shows himself to be compliant with authority and natural philosophy. He indicated a conviction that reestablishing the orbit of the sun was crucial for the renewal of astronomy but that, unlike Coper-

---

vant passages were indeed suggestive, but Reinhold's lack of interest in physical problems establishes Tycho's position as the first to assert a geoheliocentric model as the physical reality of the universe; see: Schofield, "Geoheliocentric Mathematical Hypothesis," 292–293, 296; Schofield Jones, *Tychonic and Semi-Tychonic World Systems*, 21–25.

143 Gingerich and Westman, "Wittich Connection."

144 Jardine and Segonds, *La guerre*; Jardine, *Birth of History and Philosophy of Science*.

145 Brahe, "De mundi aetherei recentioribus phaenomenis," 4: 156: "Terram, quam incolimus, centrum universi occupare, nulloque annuo motu, ut voluit Copernicus, convolui, cum veteribus astronomis et physicorum receptis sententiis, attestantibus idipsum sacris insuper literis, citra omne dubium statuendum censeo."

nicus (and Aristarchus of Samos), he had maintained the position of the earth in the center of the universe, in line with the most ancient astronomers and natural philosophy.[146]

His line of reasoning for wanting to maintain compatibility between natural philosophy and the Bible is rather surprising, given that Tycho in the account above had defined these as places where not to look for *a priori* knowledge about the corruptibility of the celestial realm or celestial matter. This was the only subject in Tycho's writings, moreover, in reference to which the Bible revealed knowledge. He himself did not provide any further information about the content of these Scriptural passages; apparently, he felt no need to neglect the hierarchy of sciences or to enact a disciplinary crossover.

Recent historiography explains this seemingly inconsistent move by Tycho as an insistence by him on the precedence of natural philosophy over Scripture. Kenneth Howell has compared different citations from Tycho's correspondence mentioning both physical and Biblical counterarguments. Based on the premise that Tycho always mentioned Scripture second, after natural philosophy, and a letter to Johannes Kepler giving only physical reasons (for instance, that the heaviness and denseness of the earth make it more plausible that the earth rests rather than moves), Howell argues that Tycho's opposition was fueled by his view that terrestrial motion was physically impossible. Otherwise, his position would have been grounded on a naïve literal reading of the Bible, something which he did not apply elsewhere in cosmological discussion. According to Howell, the astronomer proceeded according to an implicit hierarchy of disciplines and had recourse to Scripture only when natural philosophy did not provide a definitive outcome, as it did in the case of terrestrial immobility and centrality.[147] Granada holds roughly the same opinion, explaining that natural philosophy was always of primary importance to Tycho: his astronomical inquiry determined which Scriptural tenets possessed physical reality.[148]

However, several arguments cast doubt on the supposed primacy of natural philosophy in Tycho's thought. First, crediting ancient philosophical opinion with primacy is not consonant with his earlier criticism of Scholastic methodology: he clearly spoke in favor of a non-determinist stance in the study of heaven and showed himself to be quite distrustful of Peripatetic teaching and human reason altogether. Secondly, the assumed primacy of natural philosophy ignores that terrestrial immobility and centrality were considered to be

146 Brahe, "Astronomiae instauratae progymnasmata," 2: 14.

147 Howell, *God's Two Books*, 92–97.

148 Granada, "Tycho Brahe, Caspar Peucer," 581.

divinely revealed knowledge. Finally, preferring human opinion over divine authority seems quite inappropriate to the sixteenth-century setting, and the dialectical rules concerning authority as laid out in Melanchthon's *Erotemata dialectices*. This handbook unambiguously ordained that divine authority was paramount, and this was no less true in matters of natural philosophy.

In my view, Tycho's puzzling move can be made intelligible by allowing a skeptical reading of his position, as was previously established for Melanchthon's position on terrestrial immobility, according to which it was not human reason (natural philosophy) but Holy Scripture that was the primary authority. It was the clarity of God's Word which had enlightened the obscurity of the mind; the physical reasons were adduced because they were in line with Biblical teaching. There is an explicit passage in *Progymnasmata* that favors this position. When commenting upon the Copernican stance of English mathematician Thomas Digges (c. 1546–1595) and his attribution of an annual motion to the earth, Tycho clearly indicated that he preferred revelation to opinion in the context of a possible triple motion of the earth:

> The triple motion of the earth was not only against all physical truth but in contradiction with the authority of the Bible, which should be the principal matter.[149]

The lack of explanation on Tycho's part could be explained as the result of the standardization of these arguments against the physical claims of Copernicus. Christine Jones Schofield has already referred to the fact that Tycho's physical objections to terrestrial motion—as observed in his correspondence with Christoph Rothmann, among other writings—were identical to those of Melanchthon in 1549.[150] Within about forty years, this canonical list of both Biblical and physical arguments had become so commonplace that Tycho had no need to repeat them to his fellow astronomers. The assumption that this canon had been available to Tycho, or at least passed down to him, is not farfetched given the ties between the universities of Copenhagen and Wittenberg, and the influence of Melanchthon at Copenhagen University both in matters of theology and in matters of education.

The promotion of Scripture as a primary authority is in agreement not only with Melanchthon's position on the matter (and the underlying moderate skep-

149 Brahe, "Astronomiae instauratae progymnasmata," 3:175: "[...] utpote triplici motui obnoxium, non saltem contra omnem physicam veritatem, sed et repugnante Sacrarum Literarum autoritate, quae praecipua esse debet [...]."

150 Schofield Jones, *Tychonic and Semi-Tychonic World Systems*, 92–93.

tical epistemology) but also with Tycho's depreciation of a natural philosophy that is founded on undemonstrated assumptions and human reason. Furthermore, it acknowledges the revealed status of terrestrial mobility, and upholds the dialectic rule that divine authority overrules human authority. This interpretation is far less incongruous with Tycho's appraisal of credibility as manifested in his battle against the Peripatetics.

4.2 *"More in Agreement with the Truth"*

Did Tycho expect his new hypothesis of the system of the world to represent the true structure of the universe or physical reality? An affirmative answer would be inconsistent with his skeptical remarks on the limited and earthly understanding of humans with reference to knowledge of the efficient causes of planetary motions. It would, of course, be inconsistent too with my present contention that Tycho considered humans ignorant or uncertain about that which exceeds the visible realm. In the first place, the geoheliocentric model, like other hypotheses used in astronomical practice, saved the phenomena. Tycho published his geoheliocentric model as a cometary hypothesis rather than as a standard planetary hypothesis.[151] His model was intended to save the celestial phenomena of the course of the supralunary comets. The intersection of the spheres of the sun with Mars, Venus, and Mercury was no hindrance, since the celestial orbs were proven to be neither solid nor real. Tycho attributed an imaginary status to these hypotheses. In the preface of the *Mechanica*, this point is made clear:

> This is in fact primary, that in astronomy, many celestial observations are made over a long period of time using suitable and error-free Instruments. These are subsequently to be ordered by means of geometry with hypotheses devised to agree with continuous quantities [arithmetic] and the circular and uniform motion (that the celestial bodies strive for and exercise naturally and without interruption) [...].[152]

Hypotheses are described as fictitious constructions about the efficient causes between the axiom and the observed celestial phenomena, whose credibility is

151 Vanden Broecke, "Teratology," 8.

152 Brahe, "Astronomiae instauratae mechanica," 5: 5. "Idenim omnium primum est, ut in astronomicis observationes plurimae et diutinae idoneis et errori non obnoxiis organis caelitus capiantur: quae postea per geometriam excogitatis convenientibus hypothesibus in quantitates continuas et motum circularem ac uniformem (quem coelestia naturaliter, et absque intermissione appetunt atque exercent) digeruntur [...]."

grounded on the best fit between the phenomena and the axiom. In the eighth chapter of *De phaenomenis*, he still labeled the geoheliocentric model a "new model of the system of the world," not "the system of the world." On other occasions when he believed that he had established the truth about heaven, he did not refrain from explicitly saying so.[153]

Rather than establishing the truth about the configuration of the universe, I suggest that Tycho's primary concern was to render a trustworthy account of the planetary motions by avoiding the use of an absurd model, and adopting instead one which complied as far as possible with realist preconditions such as the axiom and the revealed motion of the earth. He criticized the Ptolemaic and Copernican hypotheses for being absurd. He found Ptolemy guilty of sinning against the first principles of astronomy by assuming planetary motions that did not revolve uniformly around their own center but around the equant point. Copernicus, on the other hand, did live up to the axiom of uniform circular planetary motions but credited the heavy body of the earth with a triple motion.

> When I saw that each of these hypotheses led to considerable absurdities, I began to reflect more deeply whether another system of hypotheses could be found, which would be entirely sound with regards to both mathematics and physics, and which would even escape theological reproaches, and at the same time be in full agreement with celestial phenomena.[154]

The inability to formulate a positive claim about the invisible causes of planetary motions resulted in a desire to arrive at a position of non-absurdity rather than of truth. In a letter to Christoph Rothmann, written on 21 June 1588, Tycho explicitly stated that his system was "more in agreement" with the truth than either that of Copernicus or that of Ptolemy:

> You will find at the outset of the eighth chapter a new hypothesis of heavenly revolutions, invented by me not such a short time ago, which for many reasons I do not doubt is more true than the ancient Ptolemaic or

153 Brahe, "De mundi aetherei recentioribus phaenomenis," 4: 158.

154 Ibid., 4: 156: "Cum (inquam) utrasque has hypotheses hoc modo non leues absurditates admittere introspexissem, caepi mecum altius recolere, anne aliqua hypothesium ratio inveniri posset, quae tum mathematice, tum etiam physice undiquaque recte constaret, neque etiam theologicas censuras subterfugeret, et simul apparentiis coelestibus totaliter satisfaceret."

> more recent Copernican hypothesis, and which I notice to agree with the phenomena to the highest degree. Examine everything in my hypothesis diligently, and I have no doubt that you will concede that nothing more pleasing or more consonant with the truth can be adduced.[155]

Full agreement with truth appeared to be beyond his reach. The above may not conclusively prove that Tycho did not consider the geoheliocentric model of the universe as ultimate physical reality, but it provides sufficient reason to at least question the historiographical certainty with which this has been asserted. Like Rheticus, Tycho advanced the geoheliocentric model of the universe as a most probable claim about the universe, based on the agreement between the celestial phenomena and the axiom, and its compatibility with Holy Scripture. This interpretation acknowledges his outlook on the limitations of the human mind (mapping only parts of physical reality), his distrust of human reason when unguided by proper methodology (the abyss), and is in line with what has previously been discussed as a Lutheran epistemology. The absence of references to an exhaustive knowledge of the celestial phenomena after death meant that the adjective "temporal" was still applicable in this case, but Tycho, like Rheticus, could also be subject still to an attitude of causal ignorance of the celestial phenomena.

---

155 Tycho Brahe, letter to Christoph Rothmann, Uraniborg, 21 June 1588, 6: 132–133: "Invenies etiam ibidem circa initium capitis octavi, novam a me non ita dudum adinuentam caelestium revolutionum hypothesin, quam veteri illa Ptolemaica et recentiore Copernicea multas ob causes veriorem esse non dubito, eamque apparentiis apprime consentaneam deprehendo. Tute diligenter omnia in hac circumspicito, nihilque concinnius, aut veritati convenientius adferri posse, non ambigo quin ultro concessurus sis."

CHAPTER 6

Caspar Peucer's Turn to Mosaic Philosophy

"The world [...] was created by God from nothing, by His word alone, so that it would be an abode of human nature, in which He himself wanted to be observed and become known," wrote Caspar Peucer in his 1551 mathematics textbook *Elementa doctrinae de circulis coelestibus, et primo motu*, where it served to supplement the usual Aristotelian definition of the cosmos, and where it constituted the first of several paragraphs dealing with the structure of the universe.[1] Statements such as these integrate Peucer within the Lutheran-Wittenbergian framework outlined here: as a co-editor of the *Chronicon Carionis*, he participated in shifting the history of the Academia Wittebergensis into the history of the true church, and traced the art of astronomy back to its founding father Adam.

The figure of Caspar Peucer is of crucial importance in the Wittenbergian contributions to the sixteenth-century Lutheran astronomical revolution, thanks to his academic positions at the university from the late 1540s until the early 1570s. His importance owes much also to his extra-academic correspondence at the court of Dessau in the 1580s, and the good understandings he established with his father-in-law Philipp Melanchthon, his teacher (and later friend) Rheticus,[2] his teacher Erasmus Reinhold, and fellow-astronomer Tycho Brahe (as a correspondent).[3] At the age of fifteen, he had arrived at Wittenberg University. Five years later, in 1545, Peucer obtained his Master of Arts degree.[4] He was instructed in mathematics by both Georg Joachim Rheticus and Erasmus Reinhold. After his studies, he remained on Wittenberg premises and married Melanchthon's youngest daughter, Magdalena (*d.*1576).

1 Peucer, *Elementa*, [C7^{v}]: "Mundus [...], condita a Deo ex nihilo, solo verbo ut sit domicilium humanae naturae, in qua ipse innotescere et conspici voluit."

2 After Rheticus's exit in Wittenberg in 1542, he and Peucer stayed in contact with one another but their correspondence has been lost. Rheticus was invited to the wedding of Peucer and Melanchthon's daughter in 1550; see: Burmeister, *Georg Joachim Rhetikus*, 1: 103, 121, 136, 158, and ibid., 3: 160–162, 182–184. In the preface to Peucer's *De dimensione terrae*, Peucer mentioned that Rheticus "whose mind is exquisite and whose knowledge of the arts is great" helped him with the mathematical demonstrations in the book; see: Peucer, *De dimensione terrae*, A3^{r-v}.

3 Mosley, *Bearing the Heavens*, 35.

4 The biographical account is mainly based on Brosseder, *Im Bann der Sterne*, 22–25; Wege, "Peucer," 7: 629.

 | DOI:10.1163/9789004387751_008

In 1550 he started to teach as a professor in lower mathematics. In 1554, upon the death of Erasmus Reinhold, he assumed the chair of higher mathematics, which he exchanged for the chair in medicine in 1559. One year later, he succeeded the then deceased Melanchthon as rector of the Leucorea. A decade later, he became the personal physician to Augustus (1526–1586), Elector of Saxony, while he simultaneously remained at the Medicine Faculty at Wittenberg. In 1574, his time at Wittenberg University came to an abrupt end due to his imprisonment by the said Elector on a charge of Crypto-Calvinism.[5] Martin Luther's death in 1546 had marked a period of theological controversies between the Philippists and the orthodox Lutherans (Gnesio-Lutherans), both at the University of Wittenberg and in the Electorate of Saxony. In the 1560s, the Philippists at Wittenberg were increasingly suspected of harboring hidden Calvinist beliefs regarding doctrinal issues such as the Lord's Supper. In spite of these theological controversies, Wittenberg university remained a stronghold of Luther's theology. The Philippists, among whom Peucer had been a central figure since the late 1560s, were under suspicion of having a hidden Calvinist agenda. The Elector and his theologians, who considered themselves to be orthodox Lutherans, reproached Peucer and the Wittenberg theologians for having an unorthodox Calvinist understanding of the communion, as demonstrated in a preface to a catechism written at Wittenberg and published in 1571.[6] Eventually, Peucer was imprisoned, together with three other so-called Philippists, for a period of twelve years and spent this time in different Saxonian towns, namely in the castle of Rochlitz, in the Pleissenburg in Leipzig, and in another place of incarceration in Zeitz. In 1586, he was released at the request of Agnes Hedwig of Anhalt (1573–1616), the Elector's new wife. Peucer did not return to the town of Wittenberg but became the court physician to John George I (1567–1618), Prince of the joint Principalities of Anhalt-Dessau. He died in Dessau in 1602.

The point of departure for the first section of this chapter is Peucer's earliest chronicled interventions into cosmological debate about the Copernican model, and jumps from there to his appraisal of Tycho's geoheliocentric model in the 1580s. Afterwards, the focus will shift to his correspondence with Tycho about the particularities of the celestial realm. Interestingly, it is these contributions outside of academia that show how he deals as a theorizing astronomer with the epistemological implications of the postlapsarian relationship between man and God. More than a lamp, Scripture will be depicted

5 For further information, see: Junghans, "Luther's Wittenberg," 33; Peterson, "Philippists," 255–262.

6 Christianson, "Tycho Brahe's German Treatise," 114.

as a source of knowledge. Peucer adopted the intellectual guise of a Mosaic philosopher, meaning that he pursued a study of heaven that built upon a literal reading of Holy Scripture.[7] Again, the Lutheran mistrust of human reason emerges as the underlying impetus for his evolving into a Mosaic philosopher, a mistrust conveyed in theological anthropological terms. Again, the astronomer found himself guided by the celestial phenomena, unable to gain an *a priori* insight into divine providence. His epistemological pessimism was paired once more with a denial of the trustworthiness of a speculative natural philosophy founded on unaided conjecture. Rather than a single practice, early modern Mosaic philosophers shared the common agenda to found an alternative philosophy based on a literal reading of Genesis, accompanied by the explicit rejection of the authority of Aristotle.[8] The Mosaic or sacred or Christian philosophy aimed at uniting philosophy and theology.[9]

1 Planetary Models

When it comes to the Copernican model, Caspar Peucer at first appears to be an exponent of the so-called Wittenberg Interpretation of Copernicus's *De revolutionibus*. His textbook *Elementa* outlined to the undergraduate students his vision on the cosmological implications of Copernicus's astronomy and in so doing, consolidated the early dissemination of the ideas of the Polish astronomer within a university context.[10] Based on observational, physical, and Scriptural evidence, he rebutted Copernicus's position on terrestrial mobility for not being a trustworthy physical claim. While praising Copernicus as a second Ptolemy, he denounced the offensive absurdity of the Polish astronomer's hypotheses, denying that they had any relation to physical reality.[11] Two sections of the *Elementa* are relevant. The first is entitled *Quod terra obtineat medium mundi locum, sitq[ue] centrum universi* and deals with the astronomical consequences and absurdities of assigning a non-central position to the earth for the rising and setting of the fixed stars, equinoxes, eclipses, shadows, and the lengthening and shortening of days.[12] The second section,

7 Granada, "Tycho Brahe, Caspar Peucer," 566–580.
8 Blair, "Mosaic Physics," 32–58; Crowther, "Sacred Philosophy," 397–428.
9 Blair, "Mosaic Physics," 32, 47–48.
10 Westman. "Melanchthon Circle," 178; Tredwell and Barker, "Copernicus' First Friends," 147.
11 Granada, "Tycho Brahe, Caspar Peucer," 567; Gingerich, "From Copernicus," 517; Peucer, *Elementa*, G^{v}.
12 Peucer, *Elementa*, G1^{v}–G3^{v}.

entitled *Quod terra in medio mundi haereat fixa et immota* lists four Scriptural arguments (Ps.104–105, Ps.19:4–6, Eccl.1:4–5, and Josh.10:12)[13] and four physical reasons against a circular or rectilinear motion of the earth. Peucer repeated almost verbatim the Scriptural, observational, and physical argumentation of Melanchthon's 1549 *Initia doctrinae physicae*.[14] He most probably used the latter textbook as a thesaurus of prevalent philosophical opinions from which to extract relevant definitions and arguments. His definition of the world, rendered at the outset of this chapter, for instance, was an abridged version of Melanchthon's definition of the world in the *Initia*.[15] This line of reasoning is further supported by the fact that Peucer did not respond to the objections raised by Copernicus against Ptolemy in *De revolutionibus*.[16]

In the view of Miguel Granada, Peucer's argumentation reveals a positive view that reason and natural philosophy were capable of reaching sure conclusions about elementary truths such as the immobility and centrality of the earth.[17] However, when read through a theological anthropological lens, his argument for physical causes need not imply a positive appraisal of reason on his part. Quite the contrary, Peucer too found himself tied to a theological anthropological *a posteriori* status, a condition of temporal causal ignorance regarding the efficient causes of celestial phenomena, with the one exception being revealed knowledge of the immobility and central position of the earth. This argument will be made, first, through his religious and intellectual kinship with Melanchthon and his repetition of the latter's argument on the matter, and secondly, through his discussion relating to his endorsement of Tycho's geoheliocentric model in 1589, almost four decades later.

First of all, Peucer's *Elementa* was a textbook used to teach astronomy to students in a university, an institutional setting that was not considered to be the proper environment to discuss novel ideas. Such ideas were not to be dis-

---

13 Psalm 104:5: "Who laid the foundations of the earth, that it should not be removed for ever."; Psalm 19:4–6: "In them hath he set a tabernacle for the sun, Which is as a bridegroom coming out of his chamber, and rejoiceth as a strong man to run a race. His going forth is from the end of the heaven, and his circuit unto the ends of it: and there is nothing hid from the heat thereof."; Ecclesiastes 1:4–5: "One generation passeth away, and another generation cometh: but the earth abideth for ever. The sun also ariseth, and the sun goeth down, and hasteth to his place where he arose."; Joshua 10:12: "Then spake Joshua to the Lord in the day when the Lord delivered up the Amorites before the children of Israel, and he said in the sight of Israel, Sun, stand thou still upon Gibeon; and thou, Moon, in the valley of Ajalon."

14 Peucer, *Elementa*, G3^{v}–G5^{r}.

15 Ibid., C7^{v}. Compare with Melanchthon, "Initia," 13: 214.

16 Lerner, Segonds, and Verdet, *Nicolas Copernic*, 1: 450.

17 Granada, "Tycho Brahe, Caspar Peucer," 576.

cussed with the student-astronomers in order not to offend or to disturb their minds.[18] Both Peucer's exposition on the Copernican model in the *Elementa* and his intervention regarding the 1572 nova demonstrate a wish to comply with the established hierarchy of arts and compatibility with Scripture, more than being proof of a positive appraisal of the mind's abilities. Secondly, it is safe to presume, given their shared theological opinions and intellectual background, that Peucer shared Philipp Melanchthon's view on the obscurity of the human mind. After all, the line of reasoning pursued by Peucer was part of a canon of counterarguments established in the *Initia*.[19] The clarity of Scripture, not a process of rational argument, was used by Melanchthon as the primary evidence of the earth's immobile and central position. The physical arguments were adduced because of their compatibility with revelation. Melanchthon's moderate skepticism about the outcome of the study of natural philosophy in terms of knowledge, his use of the argument of authority as the safest way for the natural philosopher to handle this obscurity of the mind, and his preference for Aristotelian natural philosophy based on its high level of compatibility with Scripture, further substantiate the present thesis that it was revelation, not reason, that had provided a decisive answer in the debate about terrestrial mobility. It was this very argument that Rheticus had tried to anticipate in the *Epistola de terrae motu*, aiming to undermine the clarity of Scripture by comparing a series of Scriptural fragments to prove the obscurity of God's Word in this respect.

The decisive substantiation of Peucer's 1551 arguments as the products of a pessimist or moderate skeptical epistemology, is found, however, in his letter to Tycho Brahe. In 1589, at Dessau, Caspar Peucer replied favorably to his correspondent's lengthy presentation regarding his geoheliocentric model, and voiced his strong approval of the dissemination of his correspondent's work, thereby endorsing the latter's cross-disciplinary incursions. His reception of Tycho's hypotheses echoed the theological anthropological notion of the astronomer's temporal ignorance regarding the invisible causes of the planetary motions. As such, it functions as the second argument in favor of a skeptical reading of the Wittenberg canon of Scriptural and physical arguments against terrestrial mobility.

In comparison with the transparency and readiness of his correspondent's model, Caspar Peucer deemed the hypotheses of the ancients to be labyrinthine.[20] His enthusiasm for Tycho stemmed from the agreement of the obser-

18 Peucer, *Elementa*, G1^{v}.

19 Barker, "Lutheran Contribution," 48.

20 Caspar Peucer, letter to Tycho Brahe, Dessau, 10 September 1589, 7: 185.

vations with the model. He stressed the necessity for observational practice, since "the foundation of the astronomical doctrine was the certitude of observations,"[21] and he praised Tycho accordingly for his diligent and accurate study of the celestial phenomena. By contrast with his discussion of the heliocentric model, Peucer here advanced neither Scriptural nor physical arguments. There was no problem of any incongruity with God's Word, given the central place of the immobile earth in the center of the universe. The criterion for judging the validity of the model advanced by Peucer was the fact that this was the closest fit between the geoheliocentric model and the celestial phenomena. A consensus of observations was used to test the accuracy of the novel planetary model, and therefore agreement with the phenomena was the extent to which a majority of observations agreed with the novel planetary model.[22] However, as with Rheticus's defense of the Copernican model, this experiential proof did not equal exhaustive knowledge about the constitution of the universe.

> If only I may live long enough to know in what manner the consensus of observations affirms and strengthens that new structure which you have founded with your hypotheses about the superior and admirable works of God, and by which we have grasped so much of the highest wisdom of God, we will then understand when we behold the works themselves face to face as eyewitnesses.[23]

Like Rheticus, Peucer introduced the theological anthropological condition of man into his discussion regarding the ontological reality of a planetary model, presupposing that the astronomer could not transcend his *a posteriori* position whilst abiding on earth. Until man experienced a face-to-face vision of God—Peucer used the same words as Rheticus and Melanchthon—knowledge of the divine works was restricted to the best model. Perfect knowledge of God implied perfect knowledge of His creation. The implications of Rheticus's position apply in Peucer's case as well: because of the asymmetrical relation between the human intellect and divine wisdom, astronomical hypotheses could not represent the true invisible efficient cause of celestial phenomena. It

---

21 Ibid., 7: 189: "Scis fundamentum doctrinae astronomicae esse certudinem observationum."

22 Ibid., 7: 190.

23 Ibid., 7: 189–190: "Utinam vivam tamdiu, ut cognoscam, quomodo consensus experientiis comprobet ac stabiliat illam quam hypothesibus tuis condis novam fabricam in maximis et mirandis Dei operibus, quibus quantum de summa illa Dei sapientia consecuti simus, intelligimus tunc, quando αυτοπται coram opera ipsa aspiciemus."

is highly unlikely that Peucer would not have ascribed man's intellectual deficiency to Adam's Fall. He integrated his own time into a larger eschatological timeframe chronicled in the universal history of the *Chronicon Carionis*, starting with original sin, and evidenced in his prefatory history of astronomy in the *Elementa*. The Fall, like the Flood, was a historical fact. Owing to man's inevitable *a posteriori* condition, astronomical hypotheses were but probable claims. The temporal causal ignorance of the astronomer could be lifted only by an autopsic view in heaven or by revelation (as was the case for terrestrial immobility and centrality).

Peucer described the posthumous condition of man as being that of an eyewitness who can behold the works of God face-to-face. The idea of a direct witnessing of the world after death was a common theological legacy, as was shown earlier by examples drawn from the writings of Melanchthon, Reinhold and Rheticus. In the *Epistola de terrae motu*, Rheticus explicitly pointed to the eschatological expectation of the second coming of Christ on earth, whereas Melanchthon and Reinhold seemed to refer to the posthumous condition of humans. Even though Peucer's writings about the new star in 1573 point to the same apocalyptic expectancy as that of Rheticus, the present citation refers to life after death rather than after the general Resurrection.[24] Regardless of issues of eschatological timing, Peucer clearly believes that truly knowing whether the hypothetically assumed causes of the celestial effects were indeed true was possible only after death. While on earth, the adequacy of the hypotheses was restricted to agreement with the phenomena, and with revelation on the subject of terrestrial mobility.

Peucer used the term "eyewitness" to refer to the personal beholding of the planetary motions in heaven, and thus attaining exhaustive knowledge about them, implying cognition of both the visible and invisible causes of nature. According to Gianna Pomata, the Latin term *autopsia* appeared for the first time in 1564 in Jean de Gorris's *Definitiones medicae*, where it is called a word of the Empiric sect that designated direct ocular experience or inspection.[25] Pomata suggests that its appearance in the late sixteenth century, and then chiefly in medical circles, probably filled a gap in contemporary philosophical terminology, which lacked a specific term to indicate first-hand experience. The appearance and rapid spread of the term in medical circles shows the increasing success of observation as a cognitive category.[26] Peucer's usage of the term

24 Brahe, "Astronomiae instauratae progymnasmata," 3: 120–123.

25 Pomata, "Observation Rising," 65.

26 Daston, "Empire of Observation," 65–66.

most probably stemmed from his activities as a physician. In reference to Hipparchus's examination of the fixed stars, Peucer used another Empiric term that emerged within the context of the sixteenth-century rise of observation: namely, *teresis*.[27] *Teresis*, or "watching," also appeared for the first time in 1564 in Gorris's *Definitiones medicae*, where it was also called a word of the Empiric sect. The term indicated a focused and repeated observation.[28] Peucer's use of the term autopsy underlines his perception of first-hand observation as a cognitive source. Again, the distinction between knowledge and ignorance is grafted onto the distinction between the visible and the invisible. Peucer found himself restricted to an epistemological condition of temporal underdetermination because he lacked first-hand experience of God's works. In this temporal condition, human reason could yield only a hypothetical or probable claim about the constitution of the world.

Peucer's theological anthropological notion of temporal causal ignorance substantiates the skeptical reading of his arguments against Copernicus's heliocentric model in 1551. Instead of a radical shift from an optimistic to a pessimistic appraisal of man's ability to acquire truth, his writings represent a skepticism of which the seeds are found in Lutheran epistemology and its more pessimistic view of the limited knowability of the physical world. The absence of similar theological anthropological considerations in 1551 can be explained by the fact that Peucer considered terrestrial immobility and centrality to be divinely revealed knowledge. Because the obscurity of the mind was illuminated by God's Word there was no need for rendering such reflections on the actual truth of the hypothesis in question. In the absence of revelation, however, the astronomer could proceed by the criterion of the best model alone, as his consideration of Tycho's hypotheses demonstrated. When allowing a skeptical reading of Peucer's 1551 position, it becomes clear that his skeptical position on the knowability of the physical world had been present already in 1551 but had been sidelined by revelation and expectations surrounding what was and was not admissible in a university setting. By 1589, outside of university context, Peucer had arrived at a genuine repudiation of the ways of the Peripatetic natural philosophers, favoring a natural philosophy based on study of the Bible instead.

27 Peucer, letter to Brahe, Dessau, 10 September 1589, 7: 189.

28 Pomata, "Observation Rising," 65.

2 Peucer as a Mosaic Philosopher

Peucer's awareness of the deficiency of the human mind resounded in a much more explicit fashion in his treatment of several other cosmological issues in 1589. He now showed himself to be a Mosaic philosopher, turning to the Bible for both *a priori* knowledge and *a posteriori* speculation. The clear teaching of Scripture was seen to evolve into a point of departure for considering themes that he found to be obscure in Scripture. One of the sources for this Mosaic philosophy may be identified in Melanchthon's likening of God's Word to a lamp for the Christian when studying nature (a comparison also drawn by Luther), advocating the need for a natural philosophy that is in pious agreement with Holy Scripture.

The historiographical label "Mosaic philosophy" is not a straightforward one. Recognizing Stephen Menn's argument that all new Renaissance philosophies could be labeled Mosaic in the sense that they all claimed to be compatible with Holy Writ, this chapter attempts to refine the specific nature of Peucer's engagement with Scripture.[29] A similar view has been expressed by Thomas Woolford who raises questions about the particularity of a Mosaic philosophy, arguing that such a category would fit the work of all late-sixteenth century orthodox Protestants.[30] Peucer's Mosaic philosophy aligns with three characteristics described by Ann Blair for other late-Renaissance natural philosophers whom she identified as Mosaic philosophers: these, namely, were a wish for a pious philosophy, a rejection of (Aristotle's) authority, and a recourse to the Bible to the extreme point of developing a natural philosophy based on a literal reading of the text of Genesis.[31] To these I add a fourth for the case of Peucer: namely, the mistrust of the human mind in the non-discernible domain. His more than aided speculative philosophy was his pious answer to the intellectual limitations of man and complementary refutation of unaided speculation. He developed this Mosaic philosophical approach for the causes of the celestial phenomena that were not, or were hardly, within sensible reach, like the sapphire color of heaven or like the Milky Way. If not revealed by Scripture, these causes belonged, at least temporarily, to the domain of human ignorance; the best way to proceed, in the absence of knowledge of such causes, was by means of the literal text of God's Word. To cite Miguel Granada on this subject: "divine revelation was the sole clue and road to truth in cosmological matters, at least

29 Menn, "Intellectual Setting," 1: 58.

30 Woolford, "Natural Theology," 141.

31 Blair, "Mosaic Physics," 33, 35.

on a point as remote from our knowledge and as inaccessible to us as the composition and structure of the highest spheres."[32]

2.1 *Celestial Matter*

Caspar Peucer considered the literal text of Scripture a source of knowledge about nature. In his correspondence with Tycho, he asserted that Holy Writ was perspicuous on the subject of the very thin fluid substance of heaven, and on the existence of supracelestial waters. Another point to be added to this list was the immobility and centrality of the earth, as he had already discussed in 1551. A literal reading of the Bible had to rely on the authentic Hebrew text, unblemished by translation or interference from philosophers. In the process of his argumentation, Peucer developed a line of reasoning that was very similar to Luther in his commentary on Genesis.

Scripture, and even reason and experience, convinced Peucer that the expanse of heaven did not consist of solid crystal orbs but of a very subtle and pure substance, and was surrounded by dense clouds and circumfluent waters.[33] Apparently, the mathematician had touched upon the subject of celestial matter in his now lost first letter to Tycho. The latter's reply discloses that Peucer had provided him with several Scriptural arguments that supported Tycho's dissolution of the solid heavenly spheres.[34] In a subsequent letter, Peucer added even more Biblical verses in support of his view. In reference to Job 37:18 ("Hast thou with him spread out the sky, which is strong, and as a molten looking glass"), a Biblical verse seemingly arguing in favor of solid celestial spheres, he delivered an argument of which I will now provide a brief description. His case in defense of the soft and aqueous nature of heaven builds on linguistic arguments, pointing to the original Hebrew words and their complementary meanings. Peucer initiated his Biblical exegesis by stating that the divine Word, in opposition to the philosophers, reveals that the whole expanse of the ethereal region consists of "a very thin and pure ethereal substance" that is penetrable by all rays.[35] The source of confusion, in Peucer's opinion, was an incorrect translation in the Septuagint and the Vulgate. He explained to his correspondent that *rakia*, the Hebrew word used to indicate heaven in the first chapter of Genesis, referred to the stretched out character of heaven that was made with a kindled and spreading light at creation.[36] He is referring to Gen-

32 Granada, "Tycho Brahe, Caspar Peucer," 578.
33 Peucer to Brahe, Dessau, 10 September 1589, 7: 186–187.
34 Brahe, letter to Caspar Peucer, Uraniborg, 13 September 1588, 7: 127–141.
35 Peucer, letter to Brahe, Dessau, 10 September 1589, 7: 187.
36 Ibid., 7: 186.

esis 1:6 ("And God said, Let there be a firmament in the midst of the waters, and let it divide the waters from the waters.") Peucer continued by explaining what this light was. On the fourth day (Gen.1:14), the light that was originally present in the *expansum* of heaven was collected (Gen.1:3) and held in the light of stars and other bodies.[37] In order to translate the celestial bodies, the Septuagint translators had used the Greek word *stereoma*, meaning solid bodies, from which the Latin word *firmament* was derived, a word that did not reflect the soft aqueous nature of the ethereal region. The Hebrew words demonstrate that Job's bronze heavens did not refer to the substance of which heaven was made. Peucer guessed that the idea of a bronze heaven was caused by accommodation to the then common opinion of Assyrian and Egyptian philosophers who believed heaven to consist of celestial orbs.[38] This is a clear example of Peucer's mistrust of human interferences with Holy Writ, distorting the meaning of Scriptural passages. His argumentation is advanced as consonant with God's Word due to his reliance on the original text.

A reading of Luther's commentary on the second day of creation shows that the same linguistic arguments that Peucer introduced had already surfaced some fifty years earlier. The combined argument of *rakia* and *schamaïm* convinced Luther that Job was not speaking of the substance of heavens when referring to bronze but revealed the surprising and perpetual existence of yet such a subtle celestial matter.[39] Luther explained that the Hebrew word *rakia* stressed the spherical shape of heaven, implying that heaven "was spread out like a bladder taking a round shape when blown up."[40] Like Peucer, he argued that *schamaïm* derived from *ÿ aim* which designated waters, meaning that heaven was made of water.[41] Although the context is different—Peucer was making an argument about earlier translators, while Luther was commenting upon a passage—these surprising similarities prove, on the one hand, that Peucer's positions were neither new nor controversial, and on the other, that his appropriation of the literal text of Scripture as a source of knowledge about nature was most probably a theologically accepted way of dealing with Scriptural passages about nature.

---

37 Ibid.

38 Ibid.

39 Luther, "Enarratio," 42: 18–19 (Commentary on Gen.1:6).

40 Ibid., 42: 19 (Commentary on Gen.1:6): "[…] sicut vesica suis circulari forma extenditur, cum inflatur […]."

41 Ibid., 42: 18–19 (Commentary on Gen.1:6).

2.2 *Supracelestial Waters*

Peucer argued that Holy Scripture was equally unambiguous on the subject of the aqueous nature of heaven and on the existence of supracelestial waters.[42] He added several Scriptural arguments (e.g., Gen.1:6; Gen.7:11; Isa.24:18)[43] that attest to the aqueous substance of heaven and the existence of supracelestial waters based on a literal interpretation. His reasoning is supported by the original use of the word *schamaïm* in Genesis 1:6, a Hebrew word that was translated with firmament and meant aqueous.[44] Interestingly, Rheticus equally accepted the testimony of Holy Writ on the existence of supracelestial waters based on the explicit argument that, as was the case also in the questions of the position and motion of the earth, his senses could not deceive him. Peucer and Rheticus thus shared the assumption of the primacy of Scripture in philosophical reflections in cases defined by the absence of sensible impressions.

Accepting revelation was a matter of faith. Acquiescence rather than comprehension was preferred, an attitude strongly reminiscent of Luther's plea for humbly confessing ignorance. On the subject of the sub- and supracelestial waters (Gen.1:6), Peucer asked his readers, "why should I be in any doubt, given the authority and firm faith about divine truth revealed?"[45] Holy Writ, however, disclosed knowledge only to a limited extent. If in these remote issues, no decisive answer or further clarification was provided by revelation, reason would not be any more apt to dispel the lingering ignorance.

> But Scripture has not revealed what those boundaries or firmaments are, nor those cataracts, nor those closed-off celestial waters; nor will any sharp-minded philosopher fully discover these things with conjectures; nor will this problem be any better explained than how the weight of the terrestrial globe can balance itself motionless on its supports in the center of the universe.[46]

---

42 Peucer, letter to Brahe, Dessau, 10 September 1589, 7:187.

43 Gen.7:11: "In the six hundredth year of Noah's life, in the second month, the seventeenth day of the month, the same day were all the fountains of the great deep broken up, and the windows of heaven were opened"; Isa.24:18: "And it shall come to pass, that he who fleeth from the noise of the fear shall fall into the pit; and he that cometh up out of the midst of the pit shall be taken in the snare: for the windows from on high are open, and the foundations of the earth do shake."

44 Peucer, letter to Brahe, Dessau, 10 September 1589, 7:187.

45 Ibid.: "[...] cur dubitem, intuens auctoritatem et fidem immotam divinae veritatis patefactae?"

46 Peucer, letter to Brahe, Dessau, 10 September 1589, 7:187: "Sed quae sint illa repagula aut

This argument is reminiscent of Luther's comments on the second day of creation that show how Luther framed his reading about creation in Genesis as a form of acknowledgment of his own ignorance. With respect to Genesis 1:6, Luther asked himself whether the waters below the firmament were clouds but answered his own query by concluding that even though he does not comprehend, he does concur.[47]

2.3 *Aided Speculative Philosophy*

Caspar Peucer took this revealed *a priori* knowledge one step further by speculating upon other philosophical topics that were veiled by the obscurity of Holy Writ. The literal text as such became a source of speculation. Both Ann Blair and Kathleen Crowther emphasize that Mosaic philosophers employed Holy Writ in a variety of manners, ranging from the delivery of natural philosophical commentaries on the Bible to extracting new physical principles from Scripture.[48] Peucer's hermeneutics in 1589 was an example of the latter: he used Scripture as a tool to speculate piously upon the remote issues of the sapphire color of heaven, and the transposition of the fixed stars and the Milky Way, while not succumbing to the pitfalls of the human mind.

On the subject of the sapphire color of heaven, Peucer argued that it was caused by solar rays coloring the celestial waters (Gen.1:6).[49] In reference to the Milky Way, he conjectured that this was in fact the pristine light of heaven pervading the thinner part of the clouds and of the heavenly waters (Gen.1:3).[50] Finally, for unspecified Biblical-linguistic reasons, Peucer argued with respect to the fixed stars that instead of being transported through heaven by the rotating concave sphere to which they were attached, as the philosophers assumed on the authority of Aristotle, their movement was prompted by "an impulse." Anticipating objections, Caspar Peucer referred to the incomprehensible wisdom of God:

> […] why do we, who have learned much about God's works, not sing God's praise much more for the construction of his work of unsearchable wisdom and of unsearchable power? Why do we not judge the motions of the

---

firmamenta, quae cataractae, quae clausae coelestes aquas […], nec Scriptura prodidit, nec philosophorum quisquam arguti ingenii coniecturis pervestigabit, nec magis hoc perscrutabitur quisquam, quam explicatum est a quoquam, quomodo suis sese fultibus in medio mundi libret immotum terreni globi pondus."

47 Luther, "Enarratio," 42: 20 (Commentary on Gen.1:6).

48 Blair, "Mosaic Physics," 32, 48; Crowther, "Sacred Philosophy," 425.

49 Peucer, letter to Brahe, Dessau, 10 September 1589, 7: 187.

50 Ibid., 7: 188.

> stars to be ruled by the wonderful and inscrutable divine wisdom which created them? Why should we not admit that the works of God are unfathomable?[51]

Again, the notion of the world as a divinely created and governed place entailed that the observed divine works were not fully fathomable by reason, and, hence, affirmed the limited knowability of the physical world. Ironically, some forty years earlier, in the *Elementa*, Peucer had actually argued in favor of the Aristotelian tenet of the motion of the fixed stars through heaven due to their attachment to the eighth sphere, since at that time he still believed in the reality of celestial spheres.[52] This change of views indicates that, in terms of speculation, he had traded in the books of Aristotle for the Bible. It is noteworthy that Peucer presented his Bible-based suggestions as pious speculation from an *a posteriori* perspective and explicitly refrained from unwarranted dogmatism.

> I therefore entertain many thoughts about the causes of those things which appear to us in the created sky; for instance, about the sapphire color of the sky, which is beheld in the clear sky during both the day and night, but differently, being brighter and purer during the day, and darker at night. Or about the dissimilar parts of Heaven, or about the Milky Way; but I do not bring laws and dogmas to anybody from these [thoughts], but rather I write down conjectures from the Word of God, leaving it for others to decide if they would rather have faith in them, or in the reasonings of the philosophers.[53]

Like Tycho and Rheticus, Peucer was involved in a process of the humbling of the Scholastic philosopher, based on his understanding of the deficiency of the human intellect, and of making positive *a priori* claims. As in Rheticus's

51 Peucer, letter to Brahe, Dessau, 10 May 1589, 7: 188–189: "[...] cur nos recte de Dei operibus edocti non multo magi tribueremus Deo opifici laudem in structura operis impervestigabilis sapientiae et potentiae impervestigabilis? Cur non statueremus motus stellarum admiranda et imperscrutabili sapientia divina regi, qua conditi sunt ? Cur non fateremur ανεξερευνητος esse opera Dei?"

52 Peucer, *Elementa*, C7^{v}.

53 Peucer, letter to Brahe, Dessau, 10 September 1589, 7: 187–188: "Ratiocinor inde et multa de causis eorum, quae in coelo creato nobis occurrunt, ut de coeli colore sapphirino, qui interdiu noctuque coelo sereno conspicitur, dissimiliter tamen, clarius et liquidius die, obscurius nocte, de dissimilibus coeli partibus, de via lactea, quibus nemini leges et dogmata fero, sed coniecturas recito ex Dei verbo, relinquens expendendum cuique, his utrum habere fidem an philosophorum rationibus velit."

hermeneutics of obscurity and Protestant hermeneutics in general, the individual conscience of the philosopher came into play as the ultimate judge in Biblical hermeneutics regarding unrevealed knowledge. Peucer even raised the stakes of the claim as a choice between faith or reason.

Peucer's recourse to Holy Writ as a repository of philosophical conjectures manifested his distrust in the ability of unaided speculative philosophy to generate correct positive answers on philosophical questions of this nature. He voiced his displeasure at unaided philosophical activity with disdainful terminology, employing words such as "dreams," "making up" or "talking nonsense."[54] On the color of heaven, he argued that the causes adduced by philosophers would by no means suffice.[55] Peucer elaborated that, given our ignorance in matters that were barely visible, he preferred Scripture to philosophical nonsense:

> However, I believe that the Milky Way consists of the thinner parts of the clouds or celestial waters, or 'barriers', or all of these things, through which the celestial, uncreated and eternal light penetrates or casts its rays more brightly than through the thicker and more obscure parts. I do not concur with the dreams of philosophers, who imagine that it is the chaotic and compressed light of many little stars, which our eyesight, on account of its frailty, cannot make out or discern [individually]; for I prefer to philosophize with and from Scripture itself, than to talk nonsense against it with philosophers inexperienced in divine matters (especially concerning those things on which they obviously hesitate, or are completely blinded), or to invent and take for true and certain hypotheses contrary to the heavenly truth, as Copernicus did about the mobility of the earth in imitation of the ancient [philosophers].[56]

54 Ibid., 7:188.

55 Ibid.

56 Ibid.: "Viam lacteam autem conjico esse rariores partes seu nubium, seu aquarum coelestium, seu repagulorum, vel horum omnium, per quas supernum, increatum et aeternum lumen penetrat aut pellucet radiatu evidentius, quam in densioribus et obscuribus. Nec assentior philosophorum somniis, qui illam fingunt esse confusum confertumque lumen multarum et parvarum stellarum, quas propter exilitatem prospectus oculorum consequi ac discernere nequeat; malo enim cum Scriptura et ex hac ipsa philosophari, quam cum philosophis rerum divinarum imperitis hanc contra nugari (in iis praesertim, in quibus aut haerent plane, aut penitus caecutiunt) aut comminisici et assumere pro certis ac veris hypotheses veritati coelesti adversas, quod exemplo veterum Copernicus fecit in mobilitate terrae."

In this passage, Peucer acknowledged both that man's intellect was bound to what is known by the senses or by Scripture, and that he was incapable of delivering statements about the divine, like those given by Copernicus. His reproach of Copernicus was exactly what Osiander had tried to forestall in the preface to *De revolutionibus*, and Rheticus in the *Epistola de terrae motu*. Whereas Osiander had disagreed that Copernicus had done such a thing by emphasizing that mathematical hypotheses were neither true nor probable (and no hierarchy of sciences was breached), Rheticus argued that his teacher indeed acknowledged his human condition of ignorance before divine omnipotence and had introduced a claim that, while not absolutely true, could at least be regarded as most true. Peucer held to the point of view expressed by himself and Melanchthon around 1550: Copernicus had failed to acknowledge heavenly truth. Peucer clearly believed that Copernicus had assumed the truth of something that he could not know for certain. The reproach of intellectual vice is one that underpins his whole argument, and is evident especially when considering Peucer's appeal for the recognition of the unfathomable nature of divine wisdom. In his appropriation of Scripture, Peucer met the same demands outlined by Tycho and Rheticus, namely the humble recognition of the unfathomable nature of divine wisdom, and the avoidance of unwarranted determinism.

Conclusion

A call for piety resonated in the Lutheran-Wittenbergian contributions to the sixteenth-century astronomical revolution. Georg Joachim Rheticus, Tycho Brahe, and Caspar Peucer invoked piety as a criterion in evaluating the trustworthiness of cosmological claims by themselves and others. Against the background of the religious and political troubles of their times, the ongoing discussions about the reassessment of the foundations of knowledge became articulated in theological anthropological terms. Learned piety in the study of the celestial realm involved both an acknowledgment of the human epistemological condition after the Fall and living up to the asymmetrical postlapsarian relation between God and humanity. Whereas fifteenth-century astronomer Regiomontanus had heralded the absolute certainty of the mathematical disciplines to the detriment of natural philosophy and the methodological ways of Scholastic philosophy, the sixteenth-century Wittenbergian astronomers deployed and reframed this key feature of the mathematical Renaissance in terms of man's postlapsarian condition and the ensuing relationship between man and God. The theological anthropological *a posteriori* position used to substantiate their vindications of novel theories was tailored to Martin Luther's understanding of man's epistemological abilities after the Fall of Adam. On the threshold of the so-called end of time, he had painted a dark picture of the sinful human condition. Perfect cognition had made way for a temporal metaphysical ignorance, limiting man's capacity for true knowledge about nature during the *saeculum* to the formal and material causes or the sensory discernible parts of creation. Due to his sinful condition, man by definition was almost doomed to err and be tricked by his own leprous mind.

Philipp Melanchthon and the astronomers we have here discussed walked Luther's theological anthropological tightrope of postlapsarian metaphysical ignorance. The impossibility of exhaustive knowledge required the astronomers to refrain from positive *a priori* statements about alternative planetary models or other *a priori* matters in natural philosophy, unless divinely revealed. In the absence of revelation, observation-based hypotheses were the best or most probable model, but they were not final. Confessions of ignorance were a key element of Lutheran learned piety. Man had to be continuously aware of the mind's potential for self-deception. The liberty to philosophize, to opt for a different methodology, to prefer other cognitive categories, was argued for by means of epistemological pessimism, the consequence of Adam's Fall from pristine Paradise. When vindicating non-canonical cosmological claims, these theorizing astronomers depicted themselves as earthly

 | DOI:10.1163/9789004387751_009

Adams and pious philosophers, humbly inquiring into divine creation, upholding a limited knowability of divine providence. The presumption of the mind's obscurity was deployed to think freely—or rather non-canonically—in the domain of cosmology, but simultaneously prevented the occurrence of a full-fledged disciplinary role-reversal. These findings affirm the radical nature of the positional switch made by Johannes Kepler. In his *Mysterium cosmographicum*, he explicitly reassessed the Copernican system from an *a priori* perspective, the unprecedented nature of which was fully described by his teacher, the Tübingen astronomer Michael Maestlin. The use made of the Adamic figure in sixteenth-century Lutheran astronomical discourse did not resemble the optimism documented by Peter Harrison for seventeenth-century English scientists aiming to restore Adamic knowledge. Kepler and the English scientists clearly sang a different theological anthropological tune.

So too did their contemporary Scholastic philosophers, according to Georg Joachim Rheticus, Tycho Brahe, and Caspar Peucer. Epistemic virtue and vice were called into learned play. The pessimistic appraisal of the human ability for *a priori* knowledge underpinned the accusations of *hubris* thrown at the feet of the Scholastic philosophers who failed to acknowledge the unlimited freedom of divine providence. The authority argument in natural philosophy was reinterpreted as bias and increasingly squandered its credibility. Tycho's work in particular voiced the perception that Aristotle's views had been too easily taken for truth. Astronomers had to be freed from the Aristotelian lens and to proceed piously. Lutheran learned piety was thus—next to an act of contemplating divine creation—an attribute of a humble *a posteriori* methodology. The methodology of the Scholastic natural philosophers induced theory-laden observation, not evidence-based theorizing. The successive writings of the astronomers show that the discrediting of the Scholastic or Peripatetic philosopher was a gradual process. Aristotle was losing his status as pious spokesperson of divine creation. Whereas Rheticus had still been careful to argue that Copernicus, in spite of a conflict of opinion, had proceeded based on ancient methodology, Tycho and Peucer in the late 1580s saw no reason to pull their punches, but instead pictured the Scholastic philosophers as lost in the abyss of nature, as reheating old cabbage, or stumbling in the labyrinths of the human mind. Tycho and Peucer explicitly represented the authority argument as clouding the very Book of Nature itself.

The theological anthropological removal of the weight of authority, supported by the condition of man between the Fall and Redemption, simultaneously facilitated the emancipation of the divinely authored Book of Nature and the book of Scripture as autonomous cognitive categories. They were valued as guarantors of a freedom of thought or rather, a freedom of divine omnipotence. These pious books were written by a divine and not a human author, and had to be read by means of applying one's eyes to the text. The emancipation of the Book of Nature buttressed the rise of visual observation as a cognitive category. Mosaic philosophy arose as another answer to the deceitful human mind. According to Peucer, the impious Scholastics neither recognized nor upheld these postlapsarian limits in reading Scripture or studying nature: their proceedings were tainted by hubris. Lutheran epistemological pessimism clearly contributed to the early modern erosion of the authority bias, of the validity of the argument of authority, and its continuation of canonical knowledge.

The theological anthropological focus allows new conclusions and refinements in the field of the history of early modern astronomy. The endorsement of Copernican astronomy is no longer an epistemological dividing line perceived between astronomers who believed that their reasonings were bound to *quia* demonstrations and those who believed themselves to have gained insight in the *propter quid* of the celestial phenomena. Applying a theological anthropological angle to Rheticus's support of Copernicus's sun-centered theory of the universe distinguishes realism from the best model, and this resolves the seeming inconsistencies in the *Narratio prima* and the *Epistola de terrae motu*. For Rheticus, to make a positive claim of such an ontological scope would be tantamount to restricting divine omnipotence to the limits of the human mind, a misstep for which he reproached the Scholastic philosophers: to advance the heliocentric model as the *propter quid* would imply failing to recognize the unfathomable nature of God's works. The postlapsarian condition of metaphysical ignorance reintegrates Rheticus within the Wittenberg circle. No longer the historiographical odd one out, his notion of the best model aligns him also with Louvain astronomer Gemma Frisius and his "prudent realism."

These Wittenbergian contributions to the sixteenth-century Lutheran astronomical revolution add another chapter to the early modern history of the *libertas philosophandi*. The astronomers' anti-dogmatist pleas argued for a breakaway from Aristotelian bias based on an unfavorable comparison of mathematical certainty versus probable opinion, which has been shown to occur in Regiomontanus's inaugural oration at the university of Padua in 1464. The Wittenbergian astronomers voiced their anti-authoritarianist stance in theological anthropological terms, and did so outside of the university walls: in the treatises on the Copernican model composed by Rheticus while away from Wittenberg, in the self-published texts and correspondence of Tycho Brahe, and in the letters of Caspar Peucer written at the court of Dessau in the late 1580s. Judging by the rhetorical discourse of these men in their respective writings about planetary models and the particulars of the celestial realm, the new role of the theorizing astronomer was one that could only develop outside a university setting. The auditorium and the textbooks were conduits for authoritative knowledge. The evaluation, for instance, of the Copernican model by Melanchthon and Peucer in combination with instances of Wittenberg mathematics professors regarding the celestial phenomena of the 1570s show a deep concern for compliance with the canonical hierarchy of sciences and Peripatetic cosmology, because Aristotle's teachings were considered to be the most compatible with Holy Scripture. Even so, this Protestant Scholasticism was equally built on the principle of the mistrust of the human mind. A university had a different mission to fulfill. Declamations *ex officio* by Melanchthon and the astronomers show that the responsibilities of the learned as members of the true church were, first, to shed light on the Bible—Luther had wished for a curriculum in support of a Bible-based theology; secondly, to keep philosophical and theological debate pious; and thirdly, to preserve as true scholars-protectors the transmission of a learned faith amidst apocalyptic dangers. The academic community present at Wittenberg, at least for the period between the 1530s and 1570s, identified their institution as an abode for the true church. Each discipline was credited with the ability to shed a particular light on the Gospel. In line with Genesis 1:14, astronomers were keepers of time and thus of history. The mathematical astronomer situated the true church spatially and chronologically in the unfolding of a world history that was driven by divine providence, a skill of considerable importance in a confessional culture with a strong apocalyptic slant. Astronomy, moreover, was a form of learned devotion, the practice of which could not only support but also strengthen faith. This foreshadows the development at the end of the sixteenth century, when the joint reading of Holy Scripture and the Book of Nature had become an established part of Lutheran confessional culture.

Bibliography

Primary Sources

Ad clarissimum virum D. Joannem Schonerum de libris revolutionum eruditissimi viri, et mathematici excellentissimi, reverendi D. Doctoris Nicolai Copernici Torunnaei, Canonici Varmiensis, per quendam iuvenem, mathematicae studiosum Narratio prima. Gdansk: Franz Rhode, 1540.

"Ad lectorem de hypothetibus huius operis," in *Nicolai Copernici Torinensis De revolutionibus orbium coelestium.* Nuremberg: Johannes Petreius, 1543.

Andreas Osiander, letter to Georg Joachim Rheticus, Nuremberg, 15 April 1541. In *Georg Joachim Rhetikus 1514–1574.* Volume 3, edited by Karl Heinz Burmeister, 25. Wiesbaden: Guido Pressler Verlag, 1968.

Brahe, Tycho. "Astronomiae instauratae mechanica" [1598]. In *Tychonis Brahe Dani opera omnia.* Volume 5, edited by John Louis Emil Dreyer, 2–161. Copenhagen: Lybraria Gyldendaliana, 1923.

Brahe, Tycho. "Astronomiae instauratae progymnasmata" [1592]. In *Tychonis Brahe Dani opera omnia.* Volumes 2 and 3, edited by John Louis Emil Dreyer. Copenhagen: Lybraria Gyldendaliana, 1915 and 1916.

Brahe, Tycho. "De mundi aetherei recentioribus phaenomenis" [1588]. In *Tychonis Brahe Dani opera omnia.* Volume 4, edited by John Louis Emil Dreyer, 1–367. Copenhagen: Lybraria Gyldendaliana, 1922.

Brahe, Tycho. *Exercices de l'Astronomie Rénovée.* Translated by Jean Peyroux. Paris: A. Blanchard, 1982.

Brahe, Tycho. *Livre I des Lettres Astronomiques.* Translated by Jean Peyroux. Paris: A. Blanchard, 2004.

Brahe, Tycho. *Mécanique de l'Astronomie Rénovée.* Translated by Jean Peyroux. Paris: A. Blanchard, 1978.

Brahe, Tycho. "On the Origin of the Comets, and what the ancient and modern Philosophers opine and maintain concerning them." Translated by John Robert Christianson. In John Robert Christianson, "Tycho Brahe's German Treatise on the Comet of 1577: A Study in Science and Politics." *Isis* 70 (1979): 132–140.

Brahe, Tycho. "Oratio de disciplinis mathematicis" [1574]. In *Tychonis Brahe Dani opera omnia.* Volume 1, edited by John Louis Emil Dreyer, 143–178. Copenhagen: Lybraria Gyldendaliana, 1913.

Brahe, Tycho. *Sur des Phénomènes plus récents du Monde Ethéré. Livre Second.* Translated by Jean Peyroux. Paris: A. Blanchard, 1984.

Brahe, Tycho. "Vonn der Cometten Uhrsprung was die alten vnnd neuen Philosophi inn denselben vermaint vnnd dauon zuhalten sei" [1578]. In *Tychonis Brahe Dani opera omnia*. Volume 4, edited by John Louis Emil Dreyer, 379–396. Copenhagen: Lybraria Gyldendaliana, 1922.

Chronicon Carionis. Auctum multis et veteribus et recentibus historiis, in descriptionibus regnorum et gentium antiquarum …; ab exordio mundi usque ad Carolum Quintum imperatorem a Philippo Melanthone et Casparo Peucero. Wittenberg: Iohannes Crato, 1572.

"Decanus collegii philosophici Erasmus Rheinholt Salveldensis." [1549?] In untitled collection, B3r–B4r. Wittenberg: Vitus Creutzer, 1549.

"Decanus collegii philosophici in Academia Witebergensi." [1549] In untitled collection, B5r–B5v. Wittenberg: Vitus Creutzer, 1549.

"Epistola de terrae motu," [1542?] in *G.J. Rheticus' Treatise on Holy Scripture and the Motion of the Earth with translations, annotations, commentary and additional chapters on Ramus-Rheticus and the development of the problem before 1650*, edited and translated by Reijer Hooykaas. Amsterdam: North-Holland Publishing Company, 1984.

In arithmeticen praefatio Georgii Joachimi Rhetici. Wittenberg: Joseph Klug, 1536.

Joachim Rheticus, letter to the students of the Wittenberg Arts Faculty, Wittenberg, October 1541. In *Georg Joachim Rhetikus 1514–1574*. Volume 3, edited by Karl Heinz Burmeister, 25. Wiesbaden: Guido Pressler Verlag, 1968.

Liber selectarum declamationum Philippi Melanthonis, quas conscripsit, et partim ipse in schola Witenbergensi recitavit, partim aliis recitandas exhibuit. Strasbourg: Crato Mylius, 1541.

Liber selectarum declamationum Philippi Melanthonis, quas conscripsit, et partim ipse in schola Witenbergensi recitavit, partim aliis recitandas exhibuit: Tomus primus. Strasbourg: Crato Mylius, 1544.

Liber selectarum declamationum Philippi Melanthonis, quas conscripsit, et partim ipse in schola Witenbergensi recitavit, partim aliis recitandas exhibuit: Tomus primus. Strasbourg: Samuel Emmel, 1558.

Liber selectarum declamationum Philippi Melanthonis, quas conscripsit, et partim ipse in schola Witenbergensi recitavit, partim aliis recitandas exhibuit: Tomus primus. Strasbourg: Samuel Emmel, 1569.

Luther, Martin. "Commentaire du Livre de la Génèse. Chapitres 1 à 11." [1543] Translated by René-H. Esnault. In *Martin Luther. Œuvres*, Volume 17. Geneva: Labor et Fides, 1975.

Luther, Martin. "Enarratio in 1. Cap. Genesis per reverendum Patrem dominum D. Mart. Lutherum in Schola Wittembergensi." In *D. Martin Luthers Werke. Kritische Gesamtausgabe*. Volume 42, edited by Karl Drescher. Weimar: Hermann Bölhaus Nachfolger, 1911.

Melanchthon, Philipp. "At Luther's Funeral." Translated by Christine F. Salazar. In *Philip Melanchthon. Orations on Philosophy and Education*, edited by Sachiko Kusukawa, 256–264. Cambridge: Cambridge University Press, 1999.

Melanchthon, Philipp. "Baumgartnero, filio." [1552] In *Corpus Reformatorum*. Volume 7, edited by Carolus Gottlieb Bretschneider, 1125–1128. Halle: C.A. Schwetschke et filium, 1840.

Melanchthon, Philipp. "De astronomia et geographia." [1536?] In *Corpus Reformatorum*. Volume 11, edited by Carolus Gottlieb Bretschneider, 292–298. Halle: C.A. Schwetschke et filium, 1843.

Melanchthon, Philipp. "De Iohanne Regiomontano." [1549] In *Corpus Reformatorum*. Volume 11, edited by Carolus Gottlieb Bretschneider, 817–826. Halle: C.A. Schwetschke et Filium, 1843.

Melanchthon, Philipp. "De physica." [1542] In *Corpus Reformatorum*. Volume 11, edited by Carolus Gottlieb Bretschneider, 555–559. Halle: C.A. Schwetschke et filium, 1843.

Melanchthon, Philip. *The Dialectical Questions: Erotemata Dialectices*. Translated and edited by Jeanne Fahnestock. Leiden: Brill, 2021.

Melanchthon, Philipp. "Doctissimo viro Simoni Grynaeo, Philippus Melanthon." [1535] In *Theoricae novae planetarum Georgii Purbacchii Germani ab Erasmo Reinholdo Salveldensi pluribus figures auctae, et illustratae scholiis, quibus studiosi praeparentur, ac invitentur ad lectionem ipsius Ptolemaei*, B1^{r}–B8^{v}. Wittenberg, 1542.

Melanchthon, Philipp. "Erotemata dialectices." [1547] In *Corpus Reformatorum*. Volume 13, edited by Carolus Gottlieb Bretschneider, 506–752. Halle: C.A. Schwetschke et filium, 1846.

Melanchthon, Philipp. "Hier. Baumgartnero." [1540] In *Corpus Reformatorum*. Volume 3, edited by Carolus Gottlieb Bretschneider, 907–914. Halle: C.A. Schwetschke et filium, 1836.

Melanchthon, Philipp. "In funere Lutheri." [1546] In *Corpus Reformatorum*. Volume 11, edited by Carolus Gottlieb Bretschneider, 726–734. Halle: C.A. Schwetschke et filium, 1843.

Melanchthon, Philipp. "Initia physicae doctrinae." [1549] In *Corpus Reformatorum*. Volume 13, edited by Carolus Gottlieb Bretschneider, 180–412. Halle: C.A. Schwetschke et filium, 1846.

Melanchthon, Philipp. "Ioanni Reiffenstein." [1536] In *Corpus Reformatorum*. Volume 3, edited by Carolus Gottlieb Bretschneider, 108. Halle: C.A. Schwetschke et filium, 1836.

Melanchthon, Philipp. "M. Meienburgo." [1549] In *Corpus Reformatorum*. Volume 7, edited by Carolus Gottlieb Bretschneider, 472–477. Halle: C.A. Schwetschke et filium, 1843.

Melanchthon, Philipp. "Praef. in arithmeticen." [1536] In *Corpus Reformatorum*. Volume 11, edited by Carolus Gottlieb Bretschneider, 284–292. Halle: C.A. Schwetschke et filium, 1843.

Melanchthon, Philipp. "Preface to *On the Sphere*." Translated by Christine F. Salazar. In *Philip Melanchthon. Orations on Philosophy and Education*, edited by Sachiko Kusukawa, 105–112. Cambridge: Cambridge University Press, 1999.

Melanchthon, Philipp. "Preface to the *Book on the Soul*." Translated by Christine F. Salazar. In *Philip Melanchthon. Orations on Philosophy and Education*, edited by Sachiko Kusukawa, 152–157. Cambridge: Cambridge University Press, 1999.

Melanchthon, Philipp. "Preface to the *Commentary on the Soul*." Translated by Christine F. Salazar. In *Philip Melanchthon. Orations on Philosophy and Education*, edited by Sachiko Kusukawa, 144–151. Cambridge: Cambridge University Press, 1999.

Mylius, Martin. *Chronologia scriptorum Philippi Melanchthonis*. Görlitz, 1582.

Nicolai Copernici Torinensis De revolutionibus orbium coelestium. Nuremberg: Johannes Petreius, 1543.

"On Astronomy and Geography." Translated by Christine F. Salazar. In *Philip Melanchthon. Orations on Philosophy and Education*, edited by Sachiko Kusukawa, 113–119. Cambridge: Cambridge University Press, 1999.

"On Johannes Regiomontanus." Translated by Christine F. Salazar. In *Philip Melanchthon. Orations on Philosophy and Education*, edited by Sachiko Kusukawa, 236–247. Cambridge: Cambridge University Press, 1999.

"On Natural Philosophy." Translated by Christine F. Salazar. In *Philip Melanchthon. Orations on Philosophy and Education*, edited by Sachiko Kusukawa, 133–138. Cambridge: Cambridge University Press, 1999.

"Oratio de astronomia et geographia, habita Wittebergae, a Ioachimo Rhetico," [1542?] in *Orationes duae, prima de astronomia et geographia, altera de physica, habitae Wittebergae a Ioachimo Rhetico, professore mathematum*, A3^v–A8^r. Nuremberg: Johannes Petreius, 1542.

"Oratio de Caspar Crucigero, ab Erasmo Rheinholt Salveldensis recitata, die vicesimo Augusti." [1549] In untitled collection, A2^r–B1^r. Wittenberg: Vitus Creutzer, 1549.

"Oratio de physica habita Wittebergae a Ioachimo Rhetico," [1542?] in *Orationes duae, prima de astronomia & geographia, altera de physica, habitae Wittebergae à Ioachimo Rhetico, professore mathematum*, A8^v–B3^r. Nuremberg: Johannes Petreius, 1542.

Peucer, Caspar. *De dimensione terrae et fontibus doctrinae longitudinis et latitudinis locorum*. Wittenberg: Iohannes Crato, 1550.

Peucer, Caspar. *Elementa doctrinae de circulis coelestibus, et primo motu*. Wittenberg: Iohannes Crato, 1551.

Peucer, Caspar. Letter to Wilhelm IV, Wittenberg, 1 January 1573. In *Tychonis Brahe Dani Opera Omnia*. Volume 3, edited by John Louis Emil Dreyer, 120–123. Copenhagen: Lybraria Gyldendaliana, 1916.

Peucer, Caspar. Letter to Tycho Brahe, Dessau, 10 May 1589. In *Tychonis Brahe Dani Opera Omnia*. Volume 7, edited by John Louis Emil Dreyer, 184–191. Copenhagen: Lybraria Gyldendaliana, 1924.

Peucer, Caspar. Letter to Tycho Brahe, Dessau, 10 September 1589. In *Tychonis Brahe Dani Opera Omnia*. Volume 7, edited by John Louis Emil Dreyer, 184–191. Copenhagen: Lybraria Gyldendaliana, 1924.

"Preface to arithmetic." Translated by Christine F. Salazar. In *Philip Melanchthon. Orations on Philosophy and Education*, edited by Sachiko Kusukawa, 90–97. Cambridge: Cambridge University Press, 1999.

Regiomontanus, Johannes. *Oratio introductoria in omnes scientias mathematicas Ioannis de Regiomonte, Patavii habita, cum Alfraganum publice praeligeret*. Nuremberg: Johannes Petreius, 1537.

Regiomontanus, Johannes. "Oratio Iohannis de Monteregio, quam habuit ipse Patavii in praelectione Alfragani." In untitled collection, $B4^{v}$–$C8^{r}$. Wittenberg: Vitus Creutzer, 1549.

Reinhold, Erasmus. "Oratio de Johanne Regiomontano mathematico, renunciatione gradus Magisterii philosophici, recitata ab Erasmo Rheinholt Salveldensi mathematum professore." [1549] In untitled collection, $A2^{r}$–$B1^{r}$. Wittenberg: Vitus Creutzer, 1549.

Reinhold, Erasmus. *Primus liber Tabulae directionum discentibus prima elementa astronomiae necessarius et utilissimus*. Tübingen: Ulricus Morhardus, 1554.

Reinhold, Erasmus. *Prutenicae tabulae coelestium motuum*. Tübingen: Ulricus Morhardus, 1551.

Rheticus, Georg Joachim. *Georgii Joachimi Rhetici Narratio Prima*. Translated by Henri Hugonnard-Roche and Jean-Pierre Verdet. With the assistance of Michel-Pierre Lerner and Alain Segonds. Wroclaw: Maison d'édition de l'académie polonaise des sciences, 1982.

Rheticus, Georg Joachim. "Narratio prima." Translated by Edward Rosen. In *Three Copernican Treatises: The* Commentariolus *of Copernicus, The Letter Against Werner, The* Narratio Prima *of Rheticus*, edited by Edward Rosen, 107–196. New York: Dover Publications, 2004.

"The Anonymous Treatise on [Holy Scripture and] the Motion of the Earth." Translated and edited by Reijer Hooykaas. In Reijer Hooykaas, *G.J. Rheticus' Treatise on Holy Scripture and the Motion of the Earth with translations, annotations, commentary and additional chapters on Ramus-Rheticus and the development of the problem before 1650*, 39–101. Amsterdam: North-Holland Publishing Company, 1984.

Theoricae novae planetarum Georgii Purbacchii Germani ab Erasmo Reinholdo Salveldensi pluribus figures auctae, et illustratae scholiis, quibus studiosi praeparentur, ac invitentur ad lectionem ipsius Ptolemaei. Wittenberg: Hans Lufft, 1542.

Tycho Brahe, letter to Christoph Rothmann, Uraniborg, 21 June 1588. In *Tychonis Brahe Dani Opera Omnia*. Volume 6, edited by John Louis Emil Dreyer, 132–133. Copenhagen: Lybraria Gyldendaliana, 1924.

Tycho Brahe, letter to Christoph Rothmann, Uraniborg, 17 August 1588. In *Tychonis Brahe Dani Opera Omnia*. Volume 6, edited by John Louis Emil Dreyer, 134–148. Copenhagen: Lybraria Gyldendaliana, 1924.

Tycho Brahe, letter to Caspar Peucer, Uraniborg, 13 September 1588. In *Tychonis Brahe Dani Opera Omnia*. Volume 7, edited by John Louis Emil Dreyer, 127–141. Copenhagen: Lybraria Gyldendaliana, 1924.

Tycho Brahe, letter to Caspar Peucer, Uraniborg, [June 1589/1590]. In *Tychonis Brahe Dani Opera Omnia*. Volume 7, edited by John Louis Emil Dreyer, 228–238. Copenhagen: Lybraria Gyldendaliana, 1924.

"4. Die Abhandlung De Terrae motu von Georg Joachim Rheticus." In *Receptio Copernicana: Texte zur Aufnahme der Copernicanischen Theorie. Kommentare und deutsche Übersetzungen.* BAND VIII/2, edited by Stefan Kirschner, Andreas Kühne and Felix Schmeidler, 69–90. Berlin, Munich, Boston: De Gruyter, 2015. https://doi.org/10.1515/9783110425475.69.

"7. Die Orationes duae, prima de astronomia & geographia, altera de physica von Georg Joachim Rheticus." In *Receptio Copernicana: Texte zur Aufnahme der Copernicanischen Theorie. Kommentare und deutsche Übersetzungen.* BAND VIII/2, edited by Stefan Kirschner, Andreas Kühne and Felix Schmeidler, 126–141. Berlin, Munich, Boston: De Gruyter, 2015. https://doi.org/10.1515/9783110425475.126.

"9. Der Kommentar zu De revolutionibus von Erasmus Reinhold." In *Receptio Copernicana: Texte zur Aufnahme der Copernicanischen Theorie. Kommentare und deutsche Übersetzungen.* BAND VIII/2, edited by Stefan Kirschner, Andreas Kühne and Felix Schmeidler, 157–375. Berlin, Munich, Boston: De Gruyter, 2015. https://doi.org/10.1515/9783110425475.157.

Literature

Aristotle. *The Metaphysics*. Translated and edited by Hugh Lawson-Tancred. London: The Penguin Group, 1998.

Arnold, Matthieu. "De la Création aux signes célestes de la fin du monde: le ciel dans quelques sermons de Martin Luther et de Jean Calvin." In *Nouveau Ciel—Nouvelle Terre: La révolution copernicienne dans l'Allemagne de la Réforme (1530–1630)*, edited by Miguel Angel Granada and Edouard Mehl, 19–39. Paris: Les Belles Lettres, 2009.

Almási, Gábor. "Rethinking Sixteenth-Century 'Lutheran Astronomy'." *Intellectual History Review* 24 (2014): 5–20. https://doi.org/10.1080/17496977.2013.841378.

Ashworth, William B. Jr. "Light of Reason, Light of Nature. Catholic and Protestant Metaphors of Scientific Knowledge." *Science in Context* 3 (1989): 89–107.

Austin, Kenneth. "Academic Exchanges: Letters, the Reformation and Scholarly self-fashioning." In *Scholarly Self-Fashioning and Community in the Early Modern University*, edited by Richard Kirwan, 39–58. Aldershot: Ashgate, 2013.

Barker, Peter. "Kepler's Epistemology." In *Method and Order in Renaissance Philosophy of Nature. The Aristotle Commentary Tradition*, edited by Daniel A. Di Liscia, Echkard Kessler, and Charlotte Methuen, 355–368. Aldershot: Ashgate, 1997.

Barker, Peter. "The Role of Religion in the Lutheran Response to Copernicus." In *Rethinking the Scientific Revolution*, edited by Margaret J. Osler, 59–88. Cambridge: Cambridge University Press, 2000.

Barker, Peter. "How Rothmann Changed His Mind." *Centaurus. International magazine of the history of mathematics, science, and technology* 46 (2004): 41–57.

Barker, Peter. "The Lutheran Contribution to the Astronomical Revolution: Science and Religion in the Sixteenth Century." In *Religious Values and the Rise of Science in Europe*, edited by John Hedley Brooke, and Ekmeleddin Ihsanoglu, 31–62. Istanbul: Research Centre for Islamic History, Art and Culture, 2005.

Barker, Peter, and Bernard R. Goldstein. "Realism and Instrumentalism in Sixteenth-Century Astronomy. A Reappraisal." *Perspectives on Science* 6 (1998): 232–258.

Barker, Peter, and Bernard R. Goldstein. "Theological Foundations of Kepler's Astronomy." In *Science in Theistic Contexts: Cognitive Dimensions*, edited by John Hedley Brooke, Margaret J. Osler and Jitse M. van der Meer. Special Issue, *Osiris* 16 (2001): 88–113.

Barnes, Robin Bruce. *Prophecy and Gnosis. Apocalypticism in the wake of the Lutheran Reformation*. Stanford: Stanford University Press, 1988.

Bauer, Barbara. "Gott, Welt, Mensch und Sterne in Melanchthons Initia doctrinae physicae." In *Melanchthon und das Lehrbuch des 16. Jahrhunderts. Begleitband zur Ausstellung im Kulturhistorschen Museum Rostock 25. April bis 13. Juli 1997*, edited by Jürgen Leonhardt, 149–173. Rostock: Universität Rostock, 1997.

Battles, Ford Lewis. "God Was Accommodating Himself to Human Capacity." *Interpretation. A Journal of Bible and Theology* 31 (1977): 19–38.

Baumgaertel, Friedrich. "Promise and Prophecy." In *The Encyclopedia of the Lutheran Church*, edited by Julius Bodensieck, 3: 1973–1974. Minneapolis: Augsburg Publishing House, 1965.

Bayer, Oswald. "Luther as an interpreter of Holy Scripture." In *The Cambridge Companion to Martin Luther*, edited by Donald McKim, 73–85. Cambridge: Cambridge University Press, 2003.

Bell, Theo. "Man is a Microcosmos: Adam and Eve in Luther's *Lectures on Genesis* (1535–1545)." *Concordia Theological Quarterly* 69 (2005): 159–184.

Bellucci, Dino. *Science de la Nature et Réformation. La physique au service de la Réforme dans l'Enseignement de Philippe Mélanchthon*. Rome: Edizione Vivere, 1998.

Benin, Stephen D. *The Footprints of God: Divine Accommodation in Jewish and Christian Thought*. New York: State University of New York Press, 1993.

Ben-Tov, Asaph. *Lutheran Humanists and Greek Antiquity. Melanchthonian Scholarship Between Universal History and Pedagogy*. Leiden: Brill, 2009.

Birkenmajer, Alexandre. "Le commentaire inédit d'Erasme Reinhold sur le *De Revolutionibus* de Nicholas Copernic." In *La Science au Seizième Siècle. Colloque de Royaumont 1–4 juillet 1957*, edited by Union Internationale d'Histoire et de Philosophie des Sciences, 169–178. Paris: Hermann, 1960.

Blair, Ann. "Tycho Brahe's Critique of Copernicus and the Copernican System." *Journal of the History of Ideas* 51 (1990): 355–377.

Blair, Ann. "Mosaic Physics and the Search for a Pious Natural Philosophy in the Late Renaissance." *Isis* 91 (2000): 32–58.

Blumenberg, Hans. *The Genesis of the Copernican World*. Translated by Robert M. Wallace. Cambridge, MA: MIT Press, 2000.

Blumenthal, Geoffrey. "Copernicus's Publication Strategy in the Contexts of Imperial and Papal Censorship and of Warmian Diplomatic Precedents." *Science in Context* 29 (2016): 151–178. https://doi:10.1017/S0269889716000016.

Blumenthal, Geoffrey. "Diplomacy, Patronage, and the Preface to De Revolutionibus." *Journal for the History of Astronomy* 44 (2013): 75–92. https://doi.org/10.1177/002182861304400104.

Boas Hall, Marie. *The Scientific Renaissance 1450–1630*. London: Collins, 1962.

Bodensieck, Julius, ed. *The Encyclopedia of the Lutheran Church*. 3 vols. Minneapolis, MPLS: Augsburg Publishing House, 1965.

Boner, Patrick J. "Kepler v. the Epicureans: causality, coincidence and the origins of the new star of 1604." *Journal for the History of Astronomy* 38 (2007): 207–221.

Boner, Patrick J. "Introduction." In *Change and Continuity in Early Modern Cosmology*, edited by Patrick J. Boner, 1–6. Dordrecht: Springer, 2011.

Boner, Patrick and Dario Tessicini. "Introduction." In *Celestial Novelties on the Eve of the Scientific Revolution 1540–1630*, edited by Dario Tessicini and Patrick J. Boner, VII–XVI. Florence: Leo S. Olschki Editore, 2013.

Bono, James J. "The Two Books and Adamic Knowledge: Reading the Book of Nature and Early Modern Strategies for Repairing the Effects of the Fall and of Babel." In *Nature and Scripture in the Abrahamic Religions: Up to 1700*, edited by Jitse van der Meer and Scott Mandelbrote, 299–340. Leiden: Brill, 2008.

Brecht, Martin. "Luther und die Türken." In *Europa und die Türken in der Renaissance*, edited by Bodo Guthmüller and Wilhelm Kühlmann, 9–28. Berlin, Boston: Max Niemeyer Verlag, 2000. https://doi.org/10.1515/9783110933567.9.

Brecht, Martin. "Martin Luther und Karl V." In *Aspectos históricos y culturales bajo Carlos V: Aspekte der Geschichte und Kultur unter Karl V*, edited by Christoph Strosetzki, 78–96. Frankfurt a. M., Madrid: Vervuert Verlagsgesellschaft, 2001. https://doi.org/10.31819/9783964563453-006.

Brockliss, Laurence. "Curricula." In *A History of the University in Europe. Volume II. Uni-*

versities in Early Modern Europe (1500–1800), edited by Hilde de Ridder-Symoens, 563–620. Cambridge: Cambridge University Press, 1996.

Brosseder, Claudia. *Im Bann der Sterne: Caspar Peucer, Philipp Melanchthon und andere Wittenberger Astrologen*. Berlin: Akademie Verlag, 2004.

Brosseder, Claudia. "The Writing in the Wittenberg Sky." *Journal of the History of Ideas* 66 (2005): 557–576.

Brotóns, Víctor Navarro. "Reception of Copernicus in Sixteenth-Century Spain. The Case of Diego de Zúñiga." *Isis* 86 (1995): 52–78.

Burmeister, Karl Heinz. *Georg Joachim Rhetikus 1514–1574*. 3 vols. Wiesbaden: Guido Pressler Verlag, 1967–1968.

Burnett, Amy N. "The Educational Roots of Reformed Scholasticism: Dialectic and Scriptural Exegesis in the Sixteenth Century." *Dutch Review of Church History/Nederlands Archief voor Kerkgeschiedenis* 84 (2004): 299–317.

Büttgen, Philippe. *Luther et la philosophie. Études d'histoire*. Paris: Vrin, 2011.

Callahan, James Patrick. "Claritas Scripturae: The Role of Perspicuity in Protestant Hermeneutics." *Journal of the Evangelical Theological Society* 39 (1996): 353–372.

Cabezón, José Ignacio. "Introduction." In *Scholasticism. Cross-Cultural and Comparative Perspectives*, edited by José Ignacio Cabezón, 1–18. New York: State University of New York Press, 1998.

Carroll, William E. "Thomas Aquinas on Science, *Sacra Doctrina*, and Creation." In *Nature and Scripture in the Abrahamic Religions: Up to 1700*, edited by Jitse van der Meer and Scott Mandelbrote, 219–248. Leiden: Brill, 2008.

Catto, Jeremy. "The Philosophical Context of the Renaissance Interpretation of the Bible." In *Hebrew Bible/ Old Testament. The History of its Interpretation. II: From the Renaissance to the Enlightenment*, edited by Magne Saebo, 106–122. Göttingen: Vandenhoeck & Ruprecht, 2008.

Chassagnette, Axelle. "La géométrie appliquée à la sphère terrestre. Le *De Dimensione terrae* (1550) de Caspar Peucer." *Histoire & Mesure* 21 (2006): 7–28.

Christianson, John Robert. "Tycho Brahe at the University of Copenhagen, 1559–1562." *Isis* 8 (1967): 198–203.

Christianson, John Robert. "Copernicus and the Lutherans." *The Sixteenth Century Journal* 4 (1973): 1–10.

Christianson, John Robert. "Tycho Brahe's German Treatise on the Comet of 1577: A Study in Science and Politics." *Isis* 70 (1979): 110–140.

Christianson, John Robert. *On Tycho's Island. Tycho Brahe and His Assistants 1570–1601*. Cambridge: Cambridge University Press, 2000.

Čížek, Jan. "The 'Physica Mosaica' of Johann Heinrich Alsted (1588–1638)." *Teorie vědy* 42 (2020): 117–139.

Crowther-Heyck, Kathleen. "Wonderful Secrets of Nature: Natural Knowledge and Religious Piety in Reformation Germany." *Isis* 94 (2003): 253–273.

Crowther, Kathleen M. "Sacred Philosophy, Secular Theology: The Mosaic Physics of Levinus Lemnius (1505–1568) and Francisco Vallés (1524–1592)." In *Nature and Scripture in the Abrahamic Religions: Up to 1700*, edited by Jitse van der Meer and Scott Mandelbrote, 391–428. Leiden: Brill, 2008.

Crowther, Kathleen M. *Adam and Eve in the Protestant Reformation*. Cambridge: Cambridge University Press, 2010.

Daston, Lorraine. "The Empire of Observation, 1600–1800." In *Histories of Scientific Observation*, edited by Lorraine Daston and Elizabeth Lunbeck, 81–113. Chicago: The University of Chicago Press, 2011.

Daston, Lorraine, and Elizabeth Lunbeck. "Introduction. Observation Observed." In *Histories of Scientific Observation*, edited by Lorraine Daston and Elizabeth Lunbeck, 1–9. Chicago: The University of Chicago Press, 2011.

Daston, Lorraine, and Elizabeth Lunbeck, eds. *Histories of Scientific Observation*. Chicago: The University of Chicago Press, 2011.

Davis, Grady H. "Piety." In *The Encyclopedia of the Lutheran Church*, edited by Julius Bodensieck, 3: 1906–1908. Minneapolis: Augsburg Publishing House, 1965.

Dear, Peter. "Totius in Verba. Rhetoric and Authority in the Early Royal Society." *Isis* 76 (1985): 145–161.

Dear, Peter. *Discipline & Experience: The Mathematical Way in the Scientific Revolution*. Chicago: The University of Chicago Press, 1995.

Deschauer, Stefan. *Die Arithemtik-Vorlesung des Georg Joachim Rheticus Wittenberg 1536. Eine kommentierte Edition der Handschrift x-278 (8) der Estnischen Akademischen Bibliothek*. Augsburg: Dr. Erwin Rauner Verlag, 2003.

Donahue, William H. "The Solid Planetary Spheres in Post-Copernican Natural Philosophy." In Donahue, William H. *The Dissolution of the Celestial Spheres 1595–1650*. New York: Arno Press, 1981.

Donahue, William H. "Astronomy." In *The Cambridge History of Science. Volume III: Early Modern Science*, edited by Katherine Park and Lorraine Daston, 562–595. Cambridge: Cambridge University Press, 2006.

Dreyer, John Louis Emil *Tycho Brahe; A Picture of Scientific Life and Work in the Sixteenth Century*. Edinburgh: Adam and Charles Black, 1890.

Esnault, René-H. "Introduction," in *Commentaire du Livre de la Genèse*, by Martin Luther. Translated by René-H. Esnault, 7–12. Geneva: Labor et Fides, 1975.

Fahnestock, Jeanne. "Introduction." In *Philip Melanchthon: The Dialectical Questions: Erotemata Dialectices*, edited by Jeanne Fahnestock, 1–141. Leiden: Brill, 2021.

Feldhay, Rivka. "Religion." In *The Cambridge History of Science. Volume III: Early Modern Science*, edited by Katherine Park and Lorraine Daston, 727–755. Cambridge: Cambridge University Press, 2006.

Folkerts, Menso. "Die mathematischen Studien Regiomontans in seiner Wiener Zeit." In *Regiomontanus-Studien*, edited by Günther Hamann, 175–209. Vienna: Verlag der Österreichischen Akademie der Wissenschaften, 1980.

Fraenkel, Peter. *Testimonia Patrum. The function of the Patristic argument in the theology of Philip Melanchthon*. Geneva: Librairie E. Droz, 1961.

Frank, Günter. "Luther and Melanchthon." In *Martin Luther: A Christian between Reforms and Modernity (1517–2017)*, edited by Alberto Melloni, 388–400. Berlin, Boston: De Gruyter, 2017. https://doi.org/10.1515/9783110499025-024.

Funkenstein, Amos. *Theology and the Scientific Imagination from the Middle Ages to the Seventeenth Century*. Princeton: Princeton University Press, 1986.

Gambaro, Ivana. "Geo-heliocentric Models and the Society of Jesus: from Clavius's Resistance to Dechales's Mathesis Regia." *Annals of Science* 78 (2021): 265–294. https://doi.org/10.1080/00033790.2021.1919760.

Gascoigne, John. "A reappraisal of the Role of the Universities in the Scientific Revolution." In *Reappraisals of the Scientific Revolution*, edited by David C. Lindberg and Robert S. Westman, 207–260. Cambridge: Cambridge University Press, 1990.

Gemeinhardt, Peter. "Das Chronicon Carionis und seine Überarbeitung durch Philipp Melanchthon: Der erste Verfasser: Johannes Carion (1499–1537)." In *Welt-Zeit: Christliche Weltchronistik aus zwei Jahrtausenden in Beständen der Thüringer Universitäts- und Landesbibliothek Jena*, edited by Martin Wallraff, 115–125. Berlin, New York: De Gruyter, 2005. https://doi.org/10.1515/9783110200966.115

Gingerich, Owen. "The role of Erasmus Reinhold and the Prutenic Tables in the Dissemination of Copernican Theory." In *Studia Copernicana VI/Colloquia Copernicana II. Études sur l'audience de la théorie héliocentrique. Conférences du Symposium de l'UIHPS*, edited by Jerzy Dobrzycki, 43–62. Wroclaw: Zaklad narodowy im Ossolinkskich, 1973.

Gingerich, Owen. "A Copernican perspective" In *The Nature of Scientific Discovery. A Symposium Commemorating the 500th Anniversary of the Birth of Nicolaus Copernicus*, edited by Owen Gingerich, 17–26. Washington: Smithsonian Institution Press, 1975.

Gingerich, Owen. "From Copernicus to Kepler: Heliocentrism as Model and as Reality." *Proceedings of the American Philosophical Society* 117 (1973): 513–522.

Gingerich, Owen. "Reinhold, Erasmus." In *Dictionary of Scientific Biography*, edited by A. Pitcairn and B. Rush, 11: 365–367. New York: Charles Scribner's Sons, 1975.

Gingerich, Owen, ed. *The Nature of Scientific Discovery. A Symposium Commemorating the 500th anniversary of the Birth of Nicolaus Copernicus*. Washington: Smithsonian Institution Press, 1975.

Gingerich, Owen. *An Annotated Census of Copernicus' De Revolutionibus (Nuremberg, 1543 and Basel, 1566)*. Leiden: Brill, 2002.

Gingerich, Owen. *The Book Nobody Read: Chasing the Revolutions of Nicolaus Copernicus*. New York: Walker, 2004.

Gingerich Owen, and Voelkel, James R. "Tycho Brahe's Copernican Campaign." *Journal for the History of Astronomy*, 29 (1998): 1–34.

Gingerich, Owen, and Westman, Robert S. "The Wittich Connection: Conflict and Priority in Late Sixteenth-Century Cosmology." *Transactions of the American Philosophical Society* 78 (1988): 1–148.

Goldstein, Bernard R. "Theory and Observation in Medieval Astronomy." *Isis* 63 (1972): 39–47.

Goldstein, Bernard R. "Saving the Phenomena." *Journal for the History of Astronomy* 28 (1997): 1–12.

Goldstein, Bernard R., and Barker, Peter. "The Role of Rothmann in the Dissolution in the Celestial Spheres." *The British Journal for the History of Science* 28 (1995): 385–403.

Goldstein, Bernard R., and Giora Hon. "Kepler's Move from Orbs to Orbits: Documenting a Revolutionary Scientific Concept." *Perspectives on Science* 13 (2005): 74–111.

Goulding, Robert. "Histories of Science in Early Modern Europe: Introduction." *Journal of the History of Ideas* 67 (2006): 33–40.

Grafton, Anthony. "From Apotheosis to Analysis: Some Late Renaissance Histories of Classical Astronomy." In *History and the Disciplines. The Reclassification of Knowledge in Early Modern Europe*, edited by Donald R. Kelley, 261–276. Rochester: The University of Rochester Press, 1997.

Grafton, Anthony. "Dating History: The Renaissance and Reformation of Chronology." *Daedalus* 132 (2003): 74–85.

Grafton, Anthony. *What Was History? The Art of History in Early Modern Europe*. Cambridge: Cambridge University Press, 2007.

Grafton, David D. "Martin Luther's Sources on the Turks and Islam in the midst of Ottoman imperialism." Special Issue, *The Muslim World* 107 (2017): 665–683. https://doi.org/10.1111/muwo.12215.

Granada, Miguel A. "Tycho Brahe, Caspar Peucer, and Christoph Rothmann on Cosmology and the Bible." In *Nature and Scripture in the Abrahamic Religions: Up to 1700*, edited by Jitse van der Meer and Scott Mandelbrote, 563–583. Leiden: Brill, 2008.

Granada, Miguel A., and Edouard Mehl, eds. *Nouveau Ciel—Nouvelle Terre: La révolution copernicienne dans l'Allemagne de la Réforme (1530–1630)*. Paris: Les Belles Lettres, 2009.

Granada, Miguel A. "Introduction." In *Nouveau Ciel—Nouvelle Terre: La révolution copernicienne dans l'Allemagne de la Réforme (1530–1630)*, edited by Miguel Angel Granada and Edouard Mehl, 9–18. Paris: Les Belles Lettres, 2009.

Granada, Miguel A. "Did Tycho Eliminate the Celestial Spheres Before 1586." *Journal of the History of Astronomy* 37 (2006): 125–145.

Granada, Miguel A., Adam Mosley, and Nicholas Jardine. *Christoph Rothmann's Discourse on the Comet of 1585: An Edition and Translation with Accompanying Essays*. Leiden: Brill, 2014.

Granada, Miguel A. "The Travail of Philosophical Liberty: *libertas philosophandi* in the long Sixteenth Century." *Dianoia* 31 (2020): 9–29.

Granada, Miguel A., Patrick J. Boner, and Dario Tessicini, eds. *Unifying Heaven and Earth: Essays in the History of Early Modern Cosmology*. Barcelona: Publicacions i Edicions de la Universitat de Barcelona, 2016.

Grant, Edward. "Late Medieval Thought, Copernicus, and the Scientific Revolution." *Journal for the History of Ideas* 23 (1962): 197–220.

Grant, Edward. "In Defense of the Earth's Centrality and Immobility: Scholastic Reaction to Copernicanism in the Seventeenth Century." *Transactions of the American Philosophical Society* 74 (1984): 1–69.

Grant, Edward. "Were There Significant Differences Between Medieval and Early Modern Scholastic Natural Philosophy? The Case for Cosmology." *Noûs* 18 (1984): 5–14.

Grant, Edward. *Planets, Stars, & Orbs. The Medieval Cosmos 1200–1678*. Cambridge: Cambridge University Press, 1996.

Grant, Edward. "God, Science, and Natural Philosophy in the Late Middle Ages." In *Between Demonstration and Imagination. Essays in the History of Science and Philosophy Presented to John D. North*, edited by Lodi Nauta and Arjo Vanderjagt, 243–267. Boston: Brill, 1999.

Green, Lawrence D., and James J. Murphy. *Renaissance Rhetoric Short-Title Catalogue 1460–1700. Revised and Expanded*. Aldershot: Ashgate, 2006.

Green, Lowell C. "Melanchthon's Relation to Scholasticism." In *Protestant Scholasticism. Essays in Reassessment*, edited by Carl R. Trueman and Scott R. Clark, 273–288. Carlisle: Paternoster Press, 1999.

Grendler, Paul F. "The Universities of the Renaissance and Reformation." *Renaissance Quarterly* 57 (2004): 1–42.

Gritsch, Eric W. "Luther as Bible translator." In *The Cambridge Companion to Martin Luther*, edited by Donald McKim, 62–72. Cambridge: Cambridge University Press, 2003.

Guthmüller, Bodo, and Wilhelm Kühlmann. *Europa und die Türken in der Renaissance*. Berlin, Boston: Max Niemeyer Verlag, 2000.

Hacking, Ian. *The Emergence of Probability: A Philosophical Study of Early Ideas about Probability, Induction and Statistical Inference*. Cambridge: Cambridge University Press, 1995.

Hamm, Berndt. *The Reformation of Faith in the Context of Late Medieval Theology and Piety*. Leiden: Brill, 2003. https://doi.org/10.1163/9789047402442.

Hallyn, Fernand. "L'absurdum àkroama de Copernic." *Bibliothèque d'Humanisme et Renaissance. Travaux et Documents* 62 (2000): 7–24.

Hallyn, Fernand. "Gemma Frisius: A Convinced Copernican in 1555." In "Copernicus and the Philosophy of the Copernican Revolution." Special Issue, *Filozofski vestnik* 25 (2004): 69–83.

Hallyn, Fernand. *Gemma Frisius, arpenteur de la terre et du ciel.* Paris: Honoré Champion, 2008.

Harrison, Peter. *The Bible, Protestantism and the Rise of Natural Science.* Cambridge: Cambridge University Press, 2001.

Harrison, Peter. "Original Sin and the Problem of Knowledge in Early Modern Europe." *Journal of the History of Ideas* 63 (2002): 239–259.

Harrison, Peter. "The 'Book of Nature' and Early Modern Science." In *The Book of Nature in Early Modern and Modern History*, edited by Klaas van Berkel and Arno Vanderjagt, 1–26. Leuven: Peeters, 2006.

Harrison, Peter. *The Fall of Man and the Foundations of Science.* Cambridge: Cambridge University Press, 2007.

Higgins, Shannon. "G.J. Rheticus and the Authorship of the Anonymous *Epistola de terrae motu.*" Unpublished master thesis, Dalhousie University, 2013.

Hillerbrand, Hans Joachim. "Grace." In *Historical Dictionary of the Reformation and Counterreformation*, edited by Hans J. Hillerbrand, 94–96. Lanham: The Scarecrow Press, 2000.

Hockenbery Dragseth, Jennifer, ed. *The Devil's Whore: Reason and Philosophy in the Lutheran Tradition.* Minneapolis: Fortress Press, 2011.

Hockenbery Dragseth, Jennifer. "Introduction." In *The Devil's Whore: Reason and Philosophy in the Lutheran Tradition*, edited by Jennifer Hockenbery Dragseth, 1–9. Minneapolis: Fortress Press, 2011.

Homer. *The Odyssey.* Translated by Samuel Butler. Plain Label Books, 2009.

Hooykaas, Reijer. *G.J. Rheticus' Treatise on Holy Scripture and the Motion of the Earth with translations, annotations, commentary and additional chapters on Ramus-Rheticus and the development of the problem before 1650.* Amsterdam: North Holland Publishing Company, 1984.

Howell, Kenneth J. "The Role of Biblical Interpretation in the Cosmology of Tycho Brahe." *Studies in History and Philosophy of Science* 29 (1998): 515–537.

Howell, Kenneth J. *God's Two Books: Copernican Cosmology and Biblical Interpretation in Early Modern Science.* Notre Dame, IND: University of Notre Dame Press, 2001.

Howell, Kenneth, J. "The Hermeneutics of Nature and Scripture in Early Modern science and Technology." In *Nature and Scripture in the Abrahamic Religions: Up to 1700*, edited by Jitse van der Meer and Scott Mandelbrote, 275–298. Leiden: Brill, 2008.

Huxley, G.L. "Sosigenes." In *Dictionary of Scientific Biography*, edited by Charles Coulston Gillispie, 12: 547. New York: Charles Scribner's Sons, 1975.

Imber, Colin. *The Ottoman Empire, 1300–1650. The structure of power.* Houndmills: Palgrave Macmillan, 2009.

Janz, Denis R. *Luther and Late Medieval Thomism: A Study in Theological Anthropology.* Waterloo, ONT: Wilfrid Laurier University Press, 1983.

Janz, Denis R. "Whore or Handmaid? Luther and Aquinas on the Function of Reason in Theology." In *The Devil's Whore. Reason and Philosophy in the Lutheran. Tradition*, edited by Jennifer Hockenbery Dragseth, 47–52. Minneapolis: Fortress Press, 2011.

Jardine, Nicholas. "The Forging of Modern Realism: Clavius and Kepler against the Sceptics." *Studies in History and Philosophy of Science* 10 (1979): 141–171.

Jardine, Nicholas. *The Birth of History and Philosophy of Science: Kepler's "A Defense of Tycho against Ursus" with Essays on its Provenance and Significance*. Cambridge: Cambridge University Press, 1984.

Jardine, Nicholas. "Epistemology of the Sciences." In *The Cambridge History of Renaissance Philosophy*, edited by Charles B. Schmitt et al., 685–711. Cambridge: Cambridge University Press, 2003.

Jardine, Nicholas. "How to Present a Copernican Comet: The Form and tactics of Christoph Rothmann's Dialexis on the Comet of 1585." In *Celestial Novelties on the Eve of the Scientific Revolution 1540–1630*, edited by Dario Tessicini and Patrick J. Boner, 159–184. Rome: Leo S. Olschki Editore, 2013.

Jardine, Nicholas, and Alain-Philippe Segonds. *La guerre des astronomes: la querelle au sujet de l'origine du système géo-héliocentrique à la fin du XVIe siècle*. 3 vols. Paris: Les Belles Lettres, 2008.

Jorink, Eric. *Reading the Book of Nature in the Dutch Golden Age 1575–1715*. Leiden: Brill, 2010.

Junghans, Helmar. "Luther's Wittenberg." In *The Cambridge Companion to Martin Luther*, edited by Donald McKim, 20–35. Cambridge: Cambridge University Press, 2003.

Kathe, Heinz. *Die Wittenberger Philosophische Fakultät 1502–1817*. Cologne: Böhlau Verlag, 2002.

Kirschner, Stefan, Andreas Kühne, and Felix Schmeidler, eds. *Receptio Copernicana: Texte zur Aufnahme der Copernicanischen Theorie. Kommentare und Deutsche Übersetzungen. BAND VIII/2*. Berlin: De Gruyter, 2015. https://doi.org/10.1515/9783110425475.

Kirschner, Stefan, and Andreas Kühne. "The Decline of Medieval Disputation Culture and the 'Wittenberg Interpretation of the Copernican Theory'." In *The Making of Copernicus. Early Modern Transformations of the Scientist and his Science*, edited by Wolfgang Neuber, Thomas Rahn and Claus Zittel. Leiden: Brill, 2015. https://doi.org/10.1163/9789004281127_003.

Kirwan, Richard. "Introduction. Scholarly Self-Fashioning and the Cultural History of Universities." In *Scholarly Self-Fashioning and Community in the Early Modern University*, edited by Richard Kirwan, 1–20. Aldershot: Ashgate, 2013.

Koehn, Horst. "Philipp Melanchthons Rede. Verzeichnis der im 16. Jahrhundert erschienen Drucke." *Archiv für Geschichte des Buchwesens* 25 (1984): 1277–1486.

Kenny, Anthony J.P. *Medieval Philosophy*. Oxford: Clarendon Press, 2005.

Kolb, Robert. "Luther's function in an age of confessionalization." In *The Cambridge Companion to Martin Luther*, edited by Donald McKim, 209–226. Cambridge: Cambridge University Press, 2003.

Kraai, Jesse. "Rheticus' Heliocentric Providence: A Study Concerning the Astrology, Astronomy of the Sixteenth Century." Unpublished doctoral thesis, University of Heidelberg, Heidelberg, 2000.

Kraeling, Emil G. *The Old Testament since the Reformation*. London: Lutterworth Press, 1955.

Krafft, Fritz. "Physikalische Realität oder mathematische Hypothese. Andreas Osiander und die physikalische Erneuerung der Antiken Astronomie durch Nicolaus Copernicus." *Philosophia Naturalis* 14 (1973): 243–275.

Kusukawa, Sachiko. *The Transformation of Natural Philosophy. The Case of Philip Melanchthon*. Cambridge: Cambridge University Press, 1995.

Kusukawa, Sachiko, ed. *Philip Melanchthon. Orations on Philosophy and Education*. Translated by Christine F. Salazar. Cambridge: Cambridge University Press, 1999.

Kusukawa, Sachiko. "Introduction." In *Philip Melanchthon. Orations on Philosophy and Education*, edited by Sachiko Kusukawa, xi–xxxi. Cambridge: Cambridge University Press, 1999.

Lerner, Michel-Pierre, Alain-Philippe Segonds, and Jean-Pierre Verdet, eds. *Nicolas Copernic. De revolutionibus orbium coelestium. Des révolutions des orbes célestes. Edition critique, traduction et notes. Volume I: Introduction*. Paris: Les Belles Lettres, 2015.

Lerner, Michel-Pierre. "Aux origines de la polémique anticopernicienne (I). L'Opusculum Quartum de Giovanmaria Tolosani [1547–1548]." *Revue des Sciences philosophiques et théologiques* 86 (2002): 681–721.

Lerner, Michel-Pierre. "Aux origines de la polémique anticopernicienne (II). Martin Luther, Andreas Osiander et Philipp Melanchthon." *Revue des Sciences philosophiques et théologiques* 90 (2006): 410–424.

Lerner, Michel-Pierre. "The Origin and Meaning of 'World System'." *Journal for the History of Astronomy* 36 (2005): 407–440.

Lerner, Michel-Pierre. *Le Monde des Sphères. II. La Fin du Cosmos Classique*. Paris: Les Belles Lettres, 2008.

Lotito, Mark A. *The Reformation of Historical Thought*. Leiden: Brill, 2019. https://doi.org/10.1163/9789004347953.

Lull, Timothy F. "Luther's Writings." In *The Cambridge Companion to Martin Luther*, edited by Donald McKim, 39–61. Cambridge: Cambridge University Press, 2003.

Lüthy, Christoph. "The Fourfold Democritus on the Stage of Early Modern Science." *Isis* 91 (2000): 443–479.

Maclean, Ian. "The 'Sceptical Crisis' Reconsidered: Galen, Rational Medicine and the Libertas Philosophandi." *Early Science and Medicine* 11 (2006): 247–274.

Mandelbrote, Scott, and Jim Bennett. *The Garden, the Ark, the Tower, the Temple. Biblical Metaphors of Knowledge in Early Modern Europe*. Oxford: Bodleian Library, 1998.

Malpangotto, Michela. *Regiomontano e il rinnovamento del sapere matematico e astronomico nel Quattrocento*. Bari: Cacucci Editore, 2008.

Maxfield, John A. *Luther's Lectures on Genesis and the Formation of Evangelical Identity*. Kirksville: Truman State University Press, 2008.

McGrath, Alister E. *Luther's Theology of the Cross*. Oxford: Basil Blackwell, 1985.

McGrath, Alister E. *The Intellectual Origins of the European Reformation*. Cornwall: Blackwell Publishing, 2004.

McKim, Donald, ed. *The Cambridge Companion to Martin Luther*. Cambridge: Cambridge University Press, 2003.

Mehl, Edouard. "Héliocentrisme et eschatologie: l'astronomie et le soleil de la justice'." In *Nouveau Ciel—Nouvelle Terre: La révolution copernicienne dans l'Allemagne de la Réforme (1530–1630)*, edited by Miguel Angel Granada and Edouard Mehl, 355–380. Paris: Les Belles Lettres, 2009.

Menn, Stephen. "The Intellectual Setting." In *The Cambridge History of Seventeenth-Century Philosophy. Volume 1*, edited by Daniel Garber and Michael Ayers, 33–86. Cambridge: Cambridge University Press, 2003.

Methuen, Charlotte. "Interpreting the Books of Nature and Scripture in Medieval and Early Modern Thought: An Introductory Essay." In *Nature and Scripture in the Abrahamic Religions: Up to 1700*, edited by Jitse van der Meer and Scott Mandelbrote, 179–218. Leiden: Brill, 2008.

Methuen, Charlotte. "Maestlin's Teaching of Copernicus: The Evidence of His University Textbook and Disputations." *Isis* 87 (1996): 230–247.

Methuen, Charlotte. "The Role of the Heavens in the Thought of Philip Melanchthon." *Journal of the History of Ideas* 57 (1996): 385–403.

Methuen, Charlotte. " "This Comet or New Star": Theology and the Interpretation of the Nova of 1572." *Perspectives on Science* 5 (1997): 499–515.

Methuen, Charlotte. "Special Providence and Sixteenth-Century Astronomical Observation: Some Preliminary Reflections." *Early Science and Medicine* 4 (1999): 99–113.

Methuen, Charlotte. "On the Threshold of a New Age: Expanding Horizons as the Broader Context of Scriptural Interpretation." In *Hebrew Bible/ Old Testament. The History of Its Interpretation. II: From the Renaissance to the Enlightenment*, edited by Magne Saebo, 665–690. Göttingen: Vandenhoeck & Ruprecht, 2008.

Methuen, Charlotte. "History and Heresy in the Lutheran Reformation." *Reformation & Renaissance Review* 24 (2022): 3–22. https://doi.org/10.1080/14622459.2022.2065138.

Miller, Gregory J. "The Turks." In *Martin Luther in Context*, edited by David M. Whitford, 152–159. Cambridge: Cambridge University Press, 2018. https://doi.org/10.1017/9781316596715.019.

Moran, Bruce T. "The Universe of Philip Melanchthon: Criticism and Use of the Copernican Theory." *Comitatus: A Journal of Medieval and Renaissance Studies* 4 (1973): 1–23.

Mosley, Adam. *Bearing the Heavens: Tycho Brahe and the Astronomical Community of the Late Sixteenth Century*. Cambridge: Cambridge University Press, 2007.

Müller, Nikolaus. "Zur Chronologie und Bibliographie der Reden Melanchthons (1545–1560)." In *Beiträge zur Reformationsgeschichte. Hernn Oberkonsistorialrat Professor D. Köstlin bei der Feier seines Siebzigsten Geburtstages*, edited by P. Albrecht, D. Breiger et al., 116–157 Gotha: Friedrich Andreas Perthes, 1896.

Nauert, Charles G. *Humanism and the Culture of Renaissance Europe*. Cambridge: Cambridge University Press, 2006.

Norlind, Wilhelm. "Copernicus and Luther: A Critical Study." *Isis* 44 (1953): 273–276.

Norlind, Wilhelm. "Reformation and Revolution: Copernicus' Discovery in an Era of Change." *The Nature of Scientific Discovery. A Symposium Commemorating the 500th anniversary of the Birth of Nicolaus Copernicus*, edited by Owen Gingerich, 134–169. Washington: Smithsonian Institution Press, 1975.

Nothaft, C. Philipp E. "Noah's Calendar: The Chronology of the Flood Narrative and the History of Astronomy in Sixteenth- and Seventeenth-Century Scholarship." *Journal of the Warburg and Courtauld Institutes* 74 (2011): 191–211.

Oberman, Heiko A. *Luther: mens tussen God en duivel*. Translated by C.J. den Heyer. Kampen: Uitgeversmaatschappij JH Kok, 1987.

Omodeo, Pietro Daniel. *Copernicus in the Cultural Debates of the Renaissance*. Leiden: Brill, 2014. https://doi.org/10.1163/9789004254503.

Omodeo, Pietro Daniel, and Jonathan Regier. "The Wittenberg Reception of Copernicus: At the origin of a Scholarly Tradition." In *Natural Knowledge and Aristotelianism at Early Modern Universities*, edited by Pietro Daniel Omodeo and Volkhard Wels, 83–108. Wiesbaden: Harrassowitz Verlag, 2019.

Osler, Margaret J. "Mixing Metaphors: Science and Religion or Natural Philosophy and Theology in Early Modern Europe." *History of Science* 35 (1997): 91–113.

Osler, Margaret J. "The Canonical Imperative: Rethinking the Scientific Revolution." In *Rethinking the Scientific Revolution*, edited by Margaret, J. Osler, 3–22. Cambridge: Cambridge University Press, 2000.

Osler, Margaret J., ed. *Rethinking the Scientific Revolution*. Cambridge: Cambridge University Press, 2000.

Osler, Margaret J. *Reconfiguring the World. Nature, God, and Human Understanding from the Middle Ages to Early Modern Europe*. Baltimore: The John Hopkins University Press, 2010.

Park, Katherine. "Observation in the Margins, 500–1500." In *Histories of Scientific Observation*, edited by Lorraine Daston and Elizabeth Lunbeck, 15–44. Chicago: The University of Chicago Press, 2011.

Park, Katherine, and Lorraine Daston, eds. *The Cambridge History of Science. Volume III: Early Modern Science*. Cambridge: Cambridge University Press, 2006.

Peterson, Charles. "Philipp Melanchthon. The First Lutheran Philosopher." In *The Devil's Whore. Reason and Philosophy in the Lutheran. Tradition*, edited by Jennifer Hockenbery Dragseth, 69–75. Minneapolis: Fortress Press, 2011.

Peterson, Luther D. "Philippists." In *The Oxford Encyclopedia of the Reformation*, edited by Hans J. Hillerbrand, 255–262. New York: Oxford University Press, 1996.

Pettegree, Andrew. *The Book in the Renaissance*. New Haven: Yale University Press, 2010.

Pohl, Eckhard. "Regiomontanus, der Begründer der astronomischen Tradition Nürenbergs." In *Regiomontanus-Studien*, edited by Günther Hamann, 291–299. Vienna: Verlag der Österreichischen Akademie der Wissenschaften, 1980.

Pohlig, Mathias. *Zwischen Gelehrsamkeit und konfessioneller Identitätsstiftung. Lutherische Kirchen- und Universalgeschichtsschreibung 1546–1617*. Tübingen: Mohr Siebeck, 2007.

Pomata, Gianna. "Observation Rising: Birth of an Epistemic Genre, 1500–1650." In *Histories of Scientific Observation*, edited by Lorraine Daston and Elizabeth Lunbeck, 45–80. Chicago: The University of Chicago Press, 2011.

Popkin, Richard H. *The History of Scepticism: From Erasmus to Descartes*. New York: Harper Torchbooks, 1964.

Popper, Nicholas. ""Abraham, Planter of Mathematics." Histories of Mathematics and Astrology in Early Modern Europe." *Journal of the History of Ideas* 67 (2006): 87–106.

Preus, James Samuel. *From Shadow to Promise. Old Testament Interpretation from Augustine to the Young Luther*. Cambridge, MA: The Belknap Press of Harvard University Press, 1969.

Ragep, F. Jamil, and Sally P. Ragep, eds. *Tradition, Transmission, Transformation. Proceedings of Two Conferences on Pre-modern Science held at the University of Oklahoma*. Leiden: Brill, 1996.

Reed, Annette Yoshiko. "Abraham as Chaldean scientist and father of the Jews: Josephus, *ANT*. 1.154–168, and the Greco-Roman discourse about astronomy/astrology." *Journal for the Study of Judaism* 35 (2004): 119–158.

Ribbed, Walther. "Wilhelm IV." In *Allgemeine Deutsche Biographie*, edited by Königl. Akademie der Wissenschaften, 43: 32–39. Leipzig: Verlag von Dunder & Humblot, 1898.

Rose, Paul Lawrence. *The Italian Renaissance of Mathematics: Studies on humanists and mathematicians from Petrarch to Galileo*. Geneva: Librairie Droz, 1975.

Rosen, Edward. "The Ramus-Rheticus Correspondence." *Journal of the History of Ideas* 1 (1940): 363–368.

Rosen, Edward. "Rheticus, Georg Joachim." In *Dictionary of Scientific Biography*, edited by A. Pitcairn and B. Rush, 11: 395–398. New York: Charles Scribner's Sons, 1975.

Rosen, Edward, ed. and transl. *Three Copernican Treatises: The* Commentariolus *of Copernicus, The Letter Against Werner, The* Narratio Prima *of Rheticus*. New York: Dover Publications, 2004.

Rosen, Edward. "Introduction." In *Three Copernican Treatises. The Commentariolus of Copernicus, The Letter Against Werner, The Narratio Prima of Rheticus*, edited and translated by Edward Rosen, 5–53. New York: Dover Publications, 2004.

Rosen, Edward, and Karl Heinz Burmeister. "Rheticus' earliest extant letter to Paul Eber." *Isis* 61 (1970): 384–386.

Rothman, Aviva. "From Cosmos to Confession: Kepler and the Connection Between Astronomical and Religious Truth." In *Change and Continuity in Early Modern Cosmology*, edited by Patrick Boner. Dordrecht: Springer, 2011. https://doi.org/10.1007/978-94-007-0037-6_7.

Saebo, Magne. "From the Renaissance to the Enlightenment. Aspects of the Cultural and Ideological Framework of Scriptural Interpretation." In *Hebrew Bible/ Old Testament. The History of its Interpretation. II: From the Renaissance to the Enlightenment*, edited by Magne Saebo, 21–48. Göttingen: Vandenhoeck & Ruprecht, 2008.

Saebo, Magne, ed. *Hebrew Bible/ Old Testament. The History of Its Interpretation. II: From the Renaissance to the Enlightenment*. Göttingen: Vandenhoeck & Ruprecht, 2008.

Schmeidler, Felix. "Regiomontans Wirkung in der Naturwissenschaft." In *Regiomontanus-Studien*, edited by Günther Hamann, 75–89. Vienna: Verlag der Österreichischen Akademie der Wissenschaften, 1980.

Schmitz, Leonhard. "Argonautae." In *A Dictionary of Greek and Roman Biography and Mythology*, edited by William Smith, 1: 280–282. London: I.B. Tauris, 2007.

Schmitz, Leonhard. "Cyclo'pes." In *A Dictionary of Greek and Roman Biography and Mythology*, edited by William Smith, 1: 909. London: I.B. Tauris, 2007.

Schofield, Christine. "The Geoheliocentric Mathematical Hypothesis in Sixteenth-Century Planetary Theory." *The British Journal for the History of Science* 2 (1965): 291–296.

Schofield Jones, Christine. *Tychonic and Semi-Tychonic World Systems*. New York: Arno Press, 1981.

Serjeantson, Richard W. "Testimony and Proof in Early-Modern England." *Studies in History and Philosophy of Science* 30 (1999): 195–236.

Shaw, Stanford. *History of the Ottoman Empire and Modern Turkey. Volume I: Empire of the Gazis: The Rise and Decline of the Ottoman Empire, 1280–1808*. Cambridge: Cambridge University Press, 1977.

Shank, Michael H. "The Classical Scientific Tradition in Fifteenth-Century Vienna" In *Tradition, Transmission, Transformation. Proceedings of Two Conferences on Premodern Science held at the University of Oklahoma*, edited by F. Jamil Ragep and Sally P. Ragep, 115–136. Leiden: Brill, 1996.

Shank, Michael H. "Regiomontanus on Ptolemy, Physical Orbs, and Astronomical Fictionalism: Goldsteinian Themes in the "Defense of Theon against George of Trebizond"." *Perspectives on Science* 10 (2002): 179–207.

Smith, Lesley. "Nicholas of Lyra and Old Testament Interpretation." In *Hebrew Bible/Old Testament. The History of Its Interpretation. II: From the Renaissance to the Enlightenment*, edited by Magne Saebo, 49–63. Göttingen: Vandenhoeck & Ruprecht, 2008.

Snobelen, Stephen D. "'In the Language of Man': The Hermeneutics of Accommodation in the Scientific Revolution." In *Nature and Scripture in the Abrahamic Religions: Up to 1700*, edited by Jitse van der Meer and Scott Mandelbrote, 691–732. Leiden: Brill, 2008.

Soen, Violet. "From the interim of Augsburg until the Treaty of Augsburg (1548–1555)." In *Martin Luther. A Christian between Reforms and modernity (1517–2017). Volume I*, edited by Alberto Melloni, 549–564. Berlin, Boston: De Gruyter, 2017.

Špelda, Daniel. "The Search for Antediluvian Astronomy: Sixteenth- and Seventeenth-Century Astronomers' Conceptions of the Origins of the Science." *Journal for the history of astronomy* 44 (2013): 337–362.

Špelda, Daniel. "From Closed Cycles to Infinite Progress: Early Modern Historiography of Astronomy." *History of science* 53 (2015): 209–233.

Swerdlow, Noel M. "Regiomontanus on the Critical Problems of Astronomy." In *Nature, Experiment, and the Sciences. Essays on Galileo and the History of Science in Honour of Stillman Drake*, edited by Trevore H. Levere and William R. Shea, 165–198. Dordrecht: Kluwer Academic Publishers, 1990.

Swerdlow, Noel M. "Annals of Scientific Publishing: Johannes Petreius' Letter to Rheticus." *Isis* 83 (1992): 270–274.

Swerdlow, Noel M. "Science and Humanism in the Renaissance: Regiomontanus's Oration on the Dignity and Utility of the Mathematical Sciences" In *World Changes. Thomas Kuhn and the Nature of Science*, edited by Paul Horwich, 131–168. Cambridge, MA: MIT Press, 1993.

Swerdlow, Noel M. "Astronomy in the Renaissance." In *Astronomy before the Telescope*, edited by Christopher Walker, 187–230. London: British Museum Press, 1996.

Swerdlow Noel M., and Otto Neugebauer. *Mathematical Astronomy in Copernicus's De Revolutionibus*. 2 vols. New York: Springer-Verlag, 1984.

Tancred-Lawson, Hugh. "Introduction." In *The Metaphysics*, by Aristotle. Translated and edited by Hugh Lawson-Tancred, xi–lxi. London: Penguin Books, 1998.

te Velde, Rudi A. "Natural Reason in the Summa Contra Gentiles." In *Thomas Aquinas. Contemporary Philosophical Perspectives*, edited by Brian Davies, 117–140. Oxford: Oxford University Press, 2002.

Thibodeau, Kenneth F. "Science and the Reformation: The Case of Strasbourg." *The Sixteenth Century Journal* 7 (1976): 35–50.

Thorndike, Lynn. *A History of Magic and Experimental Science. Volumes V and VI. The Sixteenth Century. Volume V.* New York: Columbia University Press, 1975.

Tiililae, Osmo. "Resurrection." In *The Encyclopedia of the Lutheran Church*, edited by Julius Bodensieck, 3: 2048–2050. Minneapolis: Augsburg Publishing House, 1965.

Tirapicos, Luis. "On the Censorship of Tycho Brahe's Books in Iberia." *Annals of Science* 77 (2020): 96–107. https://doi.org/10.1080/00033790.2020.1714298.

Thoren, Victor E. *The Lord of Uraniborg: A Biography of Tycho Brahe*. Cambridge: Cambridge University Press, 1990.

Tredwell, Katherine A. "Michael Maestlin and the Fate of the Narratio Prima." *Journal for the History of Astronomy* 35 (2004): 305–325.

Tredwell, Katherine A., and Peter Barker. "Copernicus' First Friends: Physical Copernicanism from 1543 to 1610." In "Copernicus and the Philosophy of Copernican Revolutions." Special Issue, *Filozofski vestnik* 25 (2004): 143–166.

Tullio, Gregory. "Translatio Studiorum." In *Translatio Studiorum. Ancient, Medieval and Modern Bearers of Intellectual History*, edited by Marco Sgarbi, 1–21. Leiden: Brill, 2012.

Valleriani, Matteo, Beate Federau, and Olya Nicolaeva. "The Hidden Praeceptor: How Georg Rheticus Taught Geocentric Cosmology to Europe." *Perspectives on Science* 30 (2022): 407–436. https://doi.org/10.1162/posc_a_00421.

van Berkel, Klaas, and Arno Vanderjagt, eds. *The Book of Nature in Early Modern and Modern History*. Leuven: Peeters, 2006.

Vanden Broecke, Steven. "Teratology and the publication of Tycho Brahe's new world system (1588)." *Journal for the History of Astronomy*, 37 (2006): 1–17.

van der Meer, Jitse, and Scott Mandelbrote, eds. *Nature and Scripture in the Abrahamic Religions: Up to 1700*. 2 vols. Leiden: Brill, 2008.

Van Leeuwen, Henry G. *The Problem of Certainty in English Thought 1630–1690*. The Hague: Martinus Nijhoff, 1963.

Vesel, Matjaz. "What is Revolutionary in the Copernicus' *Revolutions*." In "Copernicus and the Philosophy of Copernican Revolutions." Special issue, *Filozofski vestnik* 25 (2004): 167–186.

Weichenhan, Michael. *"Ergo perit coelum ..." Die Supernova des Jahres 1572 und die Überwindung der aristotelischen Kosmologie*. Munich: Franz Steiner Verlag, 2004.

Wege, Demokr. "Peucer, Caspar." In *Deutsche Biografische Enzyklopädie*, edited by Walther Killy and Rudolf Vierhaus, 7: 629. Munich: K.G. Saur, 1998.

Wengert, Timothy J. "Caspar Cruciger (1504–1548): The Case of the Disappearing Reformer." *The Sixteenth Century Journal* 20 (1989): 417–441.

Wengert, Timothy J. "Philip Melanchthon and Wittenberg's Reform of the Theological Curriculum." In *Church and School in Early Modern Protestantism. Studies in Honor of Richard A. Muller on the Maturation of a Theological Tradition*, edited by Jordan Ballor, David Sytsma and Jason Zuidema, 313–325. Leiden: Brill, 2013. https://doi.org/10.1163/9789004258297_004.

Wengert, Timothy J. "Melanchthon: Speaking for the Reformation." *The Expository Times* 126 (2015): 313–325. https://doi-org.kuleuven.e-bronnen.be/10.1177/0014524614541903.

Westman, Robert S., ed. *The Copernican Achievement.* Berkeley: University of California Press, 1975.

Westman, Robert S. "The Melanchthon Circle, Rheticus, and the Wittenberg Interpretation of the Copernican Theory" *Isis* 66 (1975): 165–193.

Westman, Robert S. "Three Responses to the Copernican Theory: Johannes Praetorius, Tycho Brahe, and Michael Maestlin." In *The Copernican Achievement,* edited by Robert S. Westman, 285–345. Berkeley: University of California Press, 1975.

Westman, Robert S. "The Astronomer's Role in the Sixteenth Century: A Preliminary Survey." *History of Science* 18 (1980): 104–147.

Westman, Robert S. "The Copernicans and the Churches." In *God and Nature: Historical Essays on the Encounter between Christianity and Science,* edited by David C. Lindberg and Numbers Ronald, 76–113. Berkeley: University of California Press, 1986.

Westman, Robert S. *The Copernican Question: Prognostication, skepticism, and celestial order.* Berkeley: University of California Press, 2011.

Westman, Robert S. "How Did Copernicus Become a Copernican?" *Isis* 110 (2019): 296–301.

Wilson, Catherine. "Epicureanism in Early Modern Philosophy." In *The Cambridge Companion to Epicureanism,* edited by James Warren, 266–286. Cambridge: Cambridge University Press, 2009.

Wolters, Gereon. "The Silence of the Wolves, Or, Why It Took the Holy Inquisition Seventy-Three Years to Ban Copernicanism." In *The Making of Copernicus. Early Modern Transformations of the Scientist and his Science,* edited by Wolfgang Neuber, Thomas Rahn and Claus Zittel, 42–63. Leiden: Brill, 2015.

Woolford, Thomas. "Natural Theology and Natural Philosophy in the Late Renaissance." Unpublished doctoral thesis, University of Cambridge, Cambridge, 2011.

Wriedt, Markus. "Luther's Theology." In *The Cambridge Companion to Martin Luther,* edited by Donald McKim, 86–119. Cambridge: Cambridge University Press, 2003.

Wrightsman, Bruce. "Andreas Osiander's Contribution to the Copernican Achievement." In *The Copernican Achievement,* edited by Robert S. Westman, 213–243. Berkeley: University of California Press, 1975.

Zambelli, Paola. "Introduction." In *"Astrologi hallucinati." Stars and the End of the World in Luther's Time,* edited by Paola Zambelli, 1–28. New York, Berlin: Walther de Gruyter, 1986.

Zinner, Ernst. *Geschichte und Bibliographie der Astronomischen Literatur in Deutschland zur Zeit der Renaissance.* Stuttgart: Anton Hiersemann, 1964.

Index Nominum

www.ingramcontent.com/pod-product-compliance
Lightning Source LLC
Chambersburg PA
CBHW071322230926
51849CB00002B/295
* 9 7 8 9 0 0 4 3 4 9 0 1 8 *